INTRODUCTION TO QUANTITATIVE ECOLOGY

Introduction to Quantitative Ecology

Mathematical and Statistical Modelling for Beginners

Timothy E. Essington

Professor, School of Aquatic and Fisheries Sciences;
University of Washington, USA

OXFORD

UNIVERSITY PRESS

OXFORD
UNIVERSITY PRESS

Great Clarendon Street, Oxford, OX2 6DP,
United Kingdom

Oxford University Press is a department of the University of Oxford.
It furthers the University's objective of excellence in research, scholarship,
and education by publishing worldwide. Oxford is a registered trade mark of
Oxford University Press in the UK and in certain other countries

© Timothy E. Essington 2021

The moral rights of the author have been asserted

First Edition published in 2021
Impression: 1

Published in the United States of America by Oxford University Press
198 Madison Avenue, New York, NY 10016, United States of America

British Library Cataloguing in Publication Data
Data available
Library of Congress Control Number: 2021937956
ISBN 978–0–19–284347–0 (hbk.)
ISBN 978–0–19–284348–7 (pbk.)
DOI: 10.1093/oso/9780192843470.001.0001
Printed and bound by
CPI Group (UK) Ltd, Croydon, CR0 4YY

Acknowledgements

I will be forever grateful for the outstanding teachers and mentors that I've had through the years, who not only inspired me to develop my quantitative toolbox but also provided me with the template for how to effectively teach this material. Thank you Tony Starfield, Anthony Ives, and Steve Carpenter for showing me how it is done. I will always be grateful to James Kitchell for taking a chance on a graduate student he barely knew and put me in the perfect environment to succeed. We miss you, Jim.

I also thank the many excellent graduate-student teaching assistants who helped me refine my teaching, bringing fresh perspectives and a critical eye to my course material and instructional methods: Anne Beaudreau, Bridget Ferriss, Kristin Marshall, Emma Hodgson, Laura Koehn, Pamela Moriarty, Maia Kapur, Maria Kuruvilla, and Helena McMonagle. Thank you to all of the students over the past few years whose keen eye for detail helped me remove errors and whose thoughtful contributions identified where the text needed improvement.

Finally, and most importantly, thank you Shereen for all of your love, support, and encouragement. You are, and always will be, the love of my life and a source of endless inspiration.

About This Book

For decades, quantitative ecology has been taught in two distinct silos: "mathematical ecology," (sometimes called "theoretical ecology"), and "statistical ecology." For a long time, that model of teaching worked very well. Students chose which topics they wanted to focus on, and selected courses accordingly to learn in-depth knowledge in each of these topics. Because most of these courses were not required, students generally were opting into this training: those that had prior training, confidence, or innate skills in quantitative reasoning naturally gravitated towards this material.

The field of ecology has changed to the point where this training model is no longer sufficient. The application of methods that were specialized twenty years ago are now routine and ubiquitous in ecological research and literature. Academic curricula that do not specifically incorporate training in quantitative tools and their applications in ecology, conservation, and natural resource management are not preparing students for their careers. To ensure that students can evaluate contemporary scientific literature, modern-day training in ecology needs to include treatment of mathematical and statistical ecology for all students, even (or especially) for students who are intimidated, lack confidence, or have been told that "they aren't good at math." In other words, we need a more encouraging approach than what we've been doing.

This book represents my attempt at achieving this—to provide information on both mathematical and and statistical approaches that help us better understand the natural world around us.

In attempting to achieve this goal, I faced an immediate challenge: how could I present an inviting and practical approach to quantitative ecology, one that covers both mathematical and statistical modeling, that didn't leave students overwhelmed, frustrated, and uninterested?

My solution is to provide *foundational* training in the main concepts and skills in mathematical and statistical ecology, presenting concepts in the *least technical* way possible.

Some earlier readers expressed concerns with my solution. By building foundations, not specialization, there isn't enough depth to prepare individuals to tackle the more complex problems that practitioners are likely to face. I agree 100%. I also know that a large number of students in my class will never become practitioners but instead are delighted to have become quantitatively fluent: they can read primary research and understand what was done, why, and how the findings ought to be interpreted. And I also know that no single book or course will prepare individuals new to any field to tackle the problems that practitioners face every day. Are you a geneticist after you take a semester of genetics? Are you a public health expert after you take one epidemiology class? Of course not. But you built the foundations of knowledge that allow you to pursue more advanced training. That is what I've attempted to do here. For this very reason, I provide information in tiers: the foundations level for everyone, and the advanced level for those who feel they have mastered the foundations and are ready for more technical nuance.

"Can I handle this?"

You don't need extensive quantitative background to master these methods. Generally, you need to be familiar with basic algebra and remember what a derivative is and what an integral does (though I promise to never make you solve an integral). You should remember basic probability rules, and remember what p-values and confidence intervals are. You should remember what a logarithm is, and remember that there are rules that govern their properties (even if you do not remember these rules).

What else do you need? First, a bit of patience with yourself. Everything in here is a new skill (even the conceptual material). Like all skills, the only way to master them is to do them—repeatedly. To steal an analogy from a colleague of mine, one could read a twelve-volume tome on how to do the perfect pushup. You could reread it, highlight text, and write notes, but you still wouldn't be able to do a perfect pushup. The only way to do that is to practice, over and over again. And they will be harder to do than their description would leave you to believe. "Plant feet and palms on floor with hands facing forward and forearms nearly parallel to the floor. Keeping your torso, hips and legs in alignment, use your arms to raise yourself upwards, and then descend downwards but do not allow any part of your body apart from hands and feet to touch the floor." Yet, without fail, every time someone does their first push up, something goes wrong. Their backside might stick up in the air, their arms might be at a strange angle, or their legs might bend. *Doing is always harder than it sounds.* I think it is helpful to remember this as you work through the book. Yes, it will be hard, and harder than it seems from the book's text. That is a natural part of learning.

You also need a bit of patience with me. While I've tried to use notation that is sensible and consistent, the field as a whole is not terribly consistent. For instance, in mathematical ecology, the parameter r usually means "intrinsic rate of population growth." In statistics, the parameter r is the "rate parameter describing density in a negative binomial or Poisson probability distribution." We use common Greek letters like α and β all of the time in completely different contexts in dynamic equations of ecological systems and in statistical notation. I've tried my very best to avoid using the same parameter to mean more than one thing. But, when a particular field has a convention on notation, I'll usually use that convention for sake of consistency with that field.

Find a mathematics refresher in chapter 11 (pp. 183–86).

How to use this book

Some context on how I developed this text will be useful. I developed this text after nearly twenty years of teaching undergraduates and graduates quantitative ecology. In my class, I focus on the "why" of quantitative ecology. That is, how do we learn about the natural world around us through models that live on our computers? This is the true art of modeling—learning about the real world from a deep knowledge of the model world. In other words, this book is about the process of learning about the real world through models.

Still, a common demand from students was to gain more knowledge of the "how." How do we code up models? How do we do certain calculations? How can we harness the full capacities of commonly available software packages? These were reasonable demands. So, I needed a compromise.

This book reflects that compromise. Throughout the text, I emphasize the "why" and then link to the "how." I do my best to separate the two, as they are distinct topics that require different mindsets. Some people are really skilled at taking a concrete finding from a mathematical model and turning that into an abstraction, something generalizable that forms a prediction about the real word. Those same people might be terrified by programming. Others can write beautiful, elegant code that shows the logical consequences of model assumptions but struggle to make sense of the model with respect to the real world. Both skills need to be developed.

The first part builds the foundations of constructing and analyzing mathematical models. To retain common principles, notation, and model structures, these chapters will explore the different ways one can build and ask questions about populations. As a result, many other types of models aren't given any treatment (see chapter 18 for a list of additional topics). But the benefit is that you are able to see how one can make different decisions about what to include and what to omit in your model, and how those decisions are guided by different model questions. The second part builds foundations of fitting ecological models to data to estimate parameters of our models and to use models as hypothesis testing tools. The third part is dedicated to the technical "skills." You might find it most useful to first explore the conceptual foundations in each chapter and then move ahead to specific skills associated with each chapter once you feel as though you have a good grasp of the concepts.

The last part provides a synthetic modeling exercise (chapter 17) that integrates components from all of the earlier sections, while also giving you a moment to appreciate what you've learned and to think about what you might do with this knowledge, moving forward.

Throughout the book, you'll see sections labeled "Advanced"; these sections contain more detailed explorations or explanations of topics. Feel free to skip these if you are just beginning to explore quantitative ecology, saving them for after you have mastered the foundational material. By including these advanced sections, it is my hope that this book can guide you through several stages of your development as a quantitative ecologist.

Contents

Part I

Fundamentals of Dynamic Models

1

Why Do We Model?

Convenient approximations bring you closest to comprehending the true nature of things.

—Haruki Murakami, *Hard-Boiled Wonderland and the End of the World*

Model: n. A simplification of reality.

One of the first and most important things to learn about quantitative ecology is that every one of our models is, by definition, wrong (Box 1979). They are wrong because they vastly simplify the real world. Reality is a complex messy place: cause-effect relationships are obscured by deeply contextual and scale-dependent interactions among multiple moving parts. We cannot possibly hope to represent all of reality with a series of mathematical or statistical expressions. Luckily, modelers don't intend to do this. Rather, modelers simplify reality *on purpose*, so that we can better understand it.

This is no different than other ways that humans understand the world. Right now, you're reading this book, presumably concentrating on the content within it. Are you thinking about the temperature of the room right now? Are you paying attention to the person who just walked by your office? Do you see all the clutter on your desk? Do you feel the texture of your clothes on your skin? Most of the time, you are able to filter out many elements of your surroundings to focus in on the task at hand. In other words, your brain is simplifying your perception of reality at that moment so that you're not overwhelmed by stimuli.

Human cognitive development provides another example of how simplifying the world helps us understand it. Consider parents speaking to their infant child using "baby talk." This may seem like a fun way for parents to engage with their child, but it turns out that using simplified speech patterns is important to help infants learn language. This simplified "infant-directed speech" helps infants understand that the noises coming out of their parent's mouth consists of discrete words (Thiessen et al. 2005). The baby talk serves as a simplified model of the actual language, the latter being far too complex for infants to grasp.

Finally, if you look at any pioneering piece of science, you'll find a simplified version of reality behind it. Bob Paine revolutionized our understanding of community ecology

Introduction to Quantitative Ecology: Mathematical and Statistical Modelling for Beginners. Timothy E. Essington, Oxford University Press. © Timothy E. Essington 2021.
DOI: 10.1093/oso/9780192843470.003.0001

by creating the idea of keystone species species that are relatively scarce but have outsized effects on ecosystems and communities. He came to this finding through careful observations and experiments on Tatoosh Island. Through this painstaking work, Paine produced an elegant conceptual model of how the intertidal animal community worked. The starfish *Pisaster* consumes mussels, permitting barnacles to persist. Remove *Pisaster*, and mussels take over. *Pisaster* is a keystone species whose presence dictates community structure. This is an elegant yet simple model of the key processes regulating intertidal communities. Think of all the elements that are *not* included in this view: no more than three species are considered, the specific processes that dictate recruitment and settlement of barnacles and mussels are not considered (all of which have been and continue to be active areas of research), and the effect of barnacles on mussels on *Pisaster* are absent. So why was this such a powerful model? Because by removing the elements of reality that were not crucially important to the question at hand, Paine was able to strip away the noise and reveal a few key processes (competition and predation).

Hopefully, by now, you are convinced that simplifying reality is a very natural way to learn and gain understanding. Of course, the field of ecological modeling goes beyond simplifying the real world; we take those simplifications and make them explicit in the form of equations and relationships. All of our simplifications are transparent for all to see. Because this is a normal and necessary way to understand the world around us, criticizing models as "unrealistic" is foolish, because all models are, by definition, unrealistic.

Yet, if we were to stop here, our definition would be incomplete. Any unrealistic depiction of the real world could be justified under the cover that "all models are wrong." Clearly, we need to refine the terminology in some way. Starfield (1997) claimed that we seek models that are *faithful* to reality. A model that includes photosynthetic antelope is not faithful to reality. A model that includes a linkage between plant productivity and antelope population dynamics is faithful to reality, even it if does not explicitly model the process of photosynthesis, plant physiology, the actual chewing and digestion of the plants, or how antelopes allocate nutrients and energy.

A second and critical refinement is to clarify that models exist for a specific purpose. Models are not a substitute for experience (Walters 1986); rather, they provide a tool to guide our experience in very specific ways. Thus, we land on a much-improved definition for model:

model: *n*. A purposeful and faithful simplification of reality.

The term *purposeful* is really important. It is the word that allows us to make decisions about what aspects of the real world to include and which to omit. It is also the word that allows us to evaluate a model. Take, for example, the two maps in Figure 1.1. At the top is a map showing the predicted risk of large fires. On the bottom, the same area is depicted but showing main transit routes. A map, like a model, is a purposeful simplification of reality. No map contains every piece of spatial information that one might ever need. So, we evaluate the model on the basis of how well it serves its purpose. It would be foolish

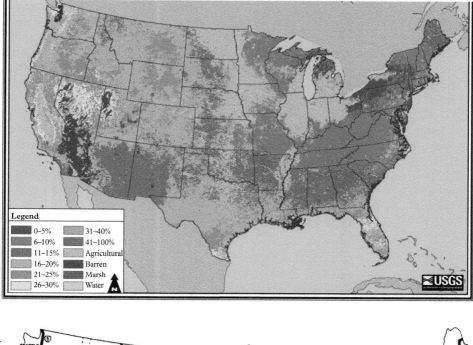

Figure 1.1 *Two maps of the United States. The top shows the probabilities of large fires in July 2017; the bottom shows the interstate highway system. Top map from the U.S. Geological Survey: https://firedanger.cr.usgs.gov/; bottom map from the U.S. Department of Transportation: https://www.fhwa.dot.gov/interstate/finalmap.cfm.*

to criticize the model at the top for failing to provide information on how to travel from Houston, TX, to Sacramento, CA. Likewise, no one would try to use the model on the bottom to deploy firefighting resources.

Because models are simplifications of reality, it is helpful to keep the boundary between the real world and the model world in mind. The real world contains everything happening at every scale. It's a terrifying place to try to do science. The model world is carved out of this place; it contains a subset of the real world. Because the model world is simple, you can understand every facet of it. In a model world, you are omniscient. You know every rule, and every relationship. Better yet, you can perform experiments on a model that you could never do in the real world to clarify cause-effect relationships.

But remember that the goal of modeling isn't to understand the model; it is to understand the real world. We want to take deep understanding of the model world and use that to answer questions about the real world. This translation between the model world and real world is impossible if you haven't carefully constructed the boundary between them:

> There is no need to ask the question "Is the model true?". If "truth" is to be the "whole truth" the answer must be "No". The only question of interest is "Is the model illuminating and useful?" (Box 1979, 203)

One last comment on the real world–model world distinction is that, while there are many excellent mathematicians and statisticians out there who can do wondrous things with equations, only a subset of them are excellent modelers. What sets them apart? They are skilled at translating between model worlds and real worlds. Note that this skill is not restricted to mathematicians and statisticians. In my experience, the ability to translate between model worlds and real worlds is the single most important skill in ecological modeling. And because model translation is a skill that anyone can learn, advanced mathematical and statistical training is not a prerequisite to becoming an effective ecological modeler. If you find yourself stuck in a mathematical quagmire, find a colleague who can help you out. In my experience, algebra gets about 60% of the job done, and calculus and linear algebra gets another 30% of the job done, leaving 10% to more advanced topics.

The value of modeling in ecology

A long-standing debate in the ecological community concerns whether more diverse food webs are more stable. The basic idea shared by many prominent ecologists in the middle of the last century was that if many species occupied similar ecological roles, the ecosystem function would be stable, because it would be less impacted by the variability in the abundance of individual species. Robert May took this seemingly

continued

intuitive idea and put it to the test mathematically: given a system of interacting species, are more complex systems likely to be more or less stable over time? The process of answering that question led to several more questions, most notably, "How do we define stable?" If we define stability as the ability of a system to quickly return to an equilibrium state after disturbance, then May showed us that more complex food webs are less stable than simple ones, in sharp contrast to the intuitive logic behind the diversity-stability hypothesis (May 1973). Did this mean that the other ecologists were wrong, or that Robert May did the math wrong? Not at all. Rather, it forced the community to think more clearly what we meant by "stability" and "complexity." Many generations of models and experimental studies have done just that: they have asked, kinds of stability exist and how do different types of food web structures promote these kinds of stability (Dunne et al. 2002)?

When you reject the extreme stances and recognize modeling as a very human way of groping for understanding, it should be obvious who will benefit most from it: those who engage in it directly. (Walters 1986, 45)

One of the greatest benefits of modeling comes from the process of model development: by forcing you to lay bare your implicit notions about how a system "works," you can expose those notions to the light of day, potentially revealing logical flaws or inconsistencies. It is akin to writing a scientific paper: you've done all of the analyses, you think you understand the entire story, but once you start writing, you may realize that your interpretation critically hinges on some implicit assumption that you didn't even know you held.

As a result, most of the learning that accompanies a modeling exercise is actually quite far removed from the endproduct of the model. Say you are building a model to predict residence times of mercury in a wetland. You find your nearest mathematical modeler, ask them to do it for you, and get the answer. All you would get would be some answer (presumably a number, perhaps with error bars). Imagine instead that you build the model yourself. Now you would immediately start asking new questions. What would govern the residence time of mercury in a wetland? Where does most of the mercury reside in a wetland? What information is available to assign numerical values to model parameters? Are there any crucial data gaps that prevent you from making a precise prediction? Is the model particularly sensitive to a small handful of assumptions or parameters? Clearly, you'll learn a lot about your own preconceptions about how the wetland processes mercury. But more importantly, the process of model development may lead you to realize that there's an alternative and far more important question to be answered. If you weren't actively engaged in the modeling process, this opportunity for learning would be lost.

A central theme of this book is that anyone can build a mathematical model to answer scientific questions. Models need not be complex or intimidating to be useful, as long as you're careful about what "use" you intend.

1.1   Myths of modeling

Starfield (1997) outlines several myths about mathematical models used for decision support in wildlife and conservation biology. Below I list several of these myths and his responses. They apply specifically to so-called tactical models, which are models that are used to aid decision-making, but they can be applied to any sort of model.

The primary purpose of building models is to make predictions: True, models predict things. Atmospheric models predict the weather from one day to the next. They take a huge amount of data, put them into complex physical models, and generate predictions about weather hazards and all manner of other things. In ecology, we rarely have the ability to predict outcomes as precisely as we can predict tomorrow's weather. We use model outputs in very different ways. Models reveal "what is possible," reveal surprising consequences of relatively simple assumptions, and help decision-making by determining when and where policy strategies are likely to be effective.

A model cannot be built with incomplete understanding of the behavior of a system: We always have to make decisions about the natural world, and information is almost always limited. A model represents a single view that summarizes your understanding of the system. In cases where there are multiple plausible interpretations for how the system works, you can build multiple models to reveal the consequences of these alternative views. Besides, if you completely understand the system, why exactly are you building a model?

It is not useful to build a model if there are gaps in the data it is likely to need (so the priority is to collect data): How do we know what data are needed? And how precise do those data need to be? Of course, having data is important, but it is rarely true that models are not useful in the absence of data. Models are hugely helpful in helping refine data-collection efforts by showing the data gaps that have the biggest consequences in terms of decision-making.

A model cannot be used in any way or form until it has been validated or been proven to be accurate: The real issue is that the model needs to be used in a way that is consistent with the model's purpose. If a model is designed to give a specific prediction about a very specific decision, then, of course, the model needs to show that the prediction is robust. But we use models in so many more ways than this.

A model must be as realistic as possible, accounting for all the detailed intricacies of a biological system: We know that all models are wrong. Remember, a model is a faithful and purposeful representation of reality. Given the purpose, does the model faithfully represent reality? If yes, then it is a good model.

Modeling is a process akin to mathematics; as such, it cannot be used or understood by most managers and many field biologists: This is the fundamental myth that I hope this text debunks. For decades, this myth led to the separation of decision-makers from stakeholders, from quantitative scientists. The field has truly changed: now modeling for decision-making engages stakeholders and decision-makers in all steps of the model process (Plagányi et al. 2013; Fulton et al. 2014). Everyone needs to understand a model to use it.

Modeling is time-consuming and expensive, so models must be designed to answer all the questions that have been thought of, or questions that may arise in the future; the more multipurpose the model, the better the value on is getting for one's investment: If a model is a purposeful representation of reality, then how can it have multiple purposes? How can you decide what the model does and does not include if you don't know how it is going to be used? Models need not be expensive and time-consuming if they are focused on a specific problem at hand.

1.2 Types of models

There is no universally agreed-upon typology of models: if you put 100 modelers in 100 different rooms and ask them to list the types of models, you'll get 100 very different lists. This is because there are so many different dimensions upon which such a list might be made. Here I review some of the main ways that models might be distinguished from each other.

1.2.1 Organizational scales

We usually think of biological systems in terms of scales: cellular/subcellular; organismal; or groups of individuals, populations, communities, food webs, ecosystems, and so on. Models might be developed around any one of those scales. Also, some models specifically address cross-scale interactions: how processes that operate at one organizational scale constrain or enhance those operating at higher or lower scales.

1.2.2 Purpose

Models are built for a variety of purposes. Perhaps the most common purpose that comes to mind is prediction. This makes sense, because models are great tools for generating quantitative predictions. Yet, models are used for quite a bit more than that. They are used to explain the natural world. By building models and understanding their properties, we can understand why systems that have properties a, b, and c are prone to x, y, and z. We also use models to assist in decision-making. Say you are deciding whether to create a wilderness protection area for an endangered species, but there is a lot of uncertainty on the species' habitat needs and threats to them. You could use a model to identify the decisions that are most robust to this uncertainty. We also use models to estimate things, especially things that cannot be observed directly. For instance, John Harte's excellent book *Consider a Spherical Cow* shows an example of estimating the total number of cobblers (people who fix shoes) in the United States that is based on some assumptions about the demand for cobblers, how many hours a day an average cobbler might work, and the length of time it takes to complete a job (Harte 1985). In other words, one can estimate the "cobbler demand" in terms of shoes per year, and the total number of shoes that a cobbler can fix per year, and from that get a rough estimate of the number of cobblers that must be working.

1.2.3   Model endpoint

Models also differ on the model endpoint. Some models are built to understand dynamic behavior (in space or time). Others are built to understand steady-state properties and how they depend on model assumptions. Other models are built to understand the structure of natural systems. Finally, some models seek to see how system-level properties can emerge from individual-level decisions.

1.2.4   Statistical or mathematical?

Some models are statistically fit to data, and others are not. Some statistical models have almost no underlying biological basis; think of the standard statistical tests you learned in your statistical-inference coursework. A linear regression (fitting a line to predict a dependent variable from an independent variable) has no biology in it but uses statistical properties to estimate the association between variables. As we will see in this book, we can also statistically fit ecological models. We might fit a model to a time series of population densities to estimate baseline conditions and rates of population growth or estimate extinction risk. We might fit a model to predict how population densities depend on habitat. Other models are never intended to be fit to data: many times we are interested in understanding the consequences of assumptions, or in defining conditions when a certain ecological outcome (e.g., coexistence of two competing species) is possible.

The focus of this book will be on dynamic models. These can be built at any organizational scale, and for any purpose, but always have the same basic structure. They depict the status of "state variables" through time or space, as a function of processes that are usually depicted as rates. A state variable is an entity, whose "state" we are modeling as a consequence of feedbacks with itself, other state variables, or perhaps other important processes. In a population context, the "state" might be the total number of individuals in a population, or the population density (number per unit area). For models of individuals, state variables might be things like size, energy storage, location, behavioral state, and so on. These models are among the most common types of ecological models and can be used in a variety of ways.

The components of these models are best explained through an example. Suppose you were developing a population model, and the only attribute you wanted to model was total population size. Population size would be your state variable. The rates are the processes that make the state variable change: in this case, births, deaths, immigration, and emigration.

Dynamic models contain state variables and rates. Rates cause the state variables to increase or decrease. State variables govern the magnitude of these rates.

We should mention a few points about dynamic models before we continue with the stages of model development. First, just because the models are called dynamic doesn't mean that we're necessarily interested in the dynamic behavior of the model. Rather, the term stems from the fact that we're modeling state variables as a function of rates, so that, mathematically, we represent the systems as functions describing rates of change. You might create a dynamic model but only calculate the equilibrium solution and ignore the dynamics. Second, we commonly use dynamic models as estimation tools or to optimize a decision. For instance, quantifying the residence time of mercury in a wetland is an estimation problem, but, to do it, we have to understand the rates at which mercury moves among ecosystem compartments, and the amount of mercury that is within each of these compartments. Thus, we need to set up a model with several state variables (mercury levels in several compartments) and with several rate processes (the flux of mercury among state variables).

1.3 Developing your model

What follows is a stepwise process for developing a dynamic model. Some of these steps will be explained in detail, while others will be summarized more briefly. A key point to remember from the outset is that the development process is never linear—it is cyclical and iterative. You start with step 1 and then proceed to step 2, which, in turn, might make you question your decisions in the first step. This happens throughout the process.

1.3.1 Steps in model development

The steps for model development are as follows:

1. Identifying your question
2. Defining the model world
3. Building a conceptual model
4. Translating conceptual models into math
5. Model coding and diagnostics
6. Interpreting and analyzing the model's output.

1.3.1.1   Identifying your question

A perfectly executed, mathematically elegant model is completely worthless without a good question motivating the model development. Moreover, not all questions are well suited for modeling. For instance, suppose you were interested in the pH tolerance of a common zooplankton, *Daphnia pulex*. It's hard to imagine how (or why) you would build a model to address this question, because it is so easily measurable. Set up a bunch of experimental aquariums, change the pH in a controlled fashion, and describe the mortality as a function of pH (side note: quantifying this effect from data will require you to apply a statistical model, so you're not totally free of modeling in this case). If your question can be readily answered through direct observation, then put your pen, paper, and computer aside and measure it.

The following are some good candidates for models:

- Any question that begins with "Under what conditions….": Here the focus is on describing the conditions under which a particular ecological phenomenon might be expected to occur. For instance, under what conditions can two competing species coexist?

- A question that involves measuring an unobservable quantity: These commonly arise in ecosystem-scale phenomenon where a simple measurement isn't going to give you an answer. For instance, Baird et al. (1991) built models of several estuarine ecosystems and calculated, among other things, the proportion of primary production that ended up in planktivorous and piscivorous fishes. It's hard to imagine how one could calculate this without some sort of model.

- When you need some sort of answer *fast*, before you have time to develop the tools and resources to directly test and /or measure something: Obviously, any answer that's motivated by this should have disclaimers on it, pointing out that the output is intended as a rapid, rough approximation.

- When you are unsure whether your own interpretation of ecological phenomenon is valid: You may observe a change in the variable X which seems to be related to changes in the variable Y, and you presume it is because Y directly affects some rate process z. You may want to know "Is it possible for Y to impact X through its effect on z?"

- To improve your decision-making based on some prediction about future behavior.

- When you seek to develop generalized statements or predictions about the natural world.

1.3.1.2   Defining the model world

Next comes the hard part of determining what aspects of reality you plan on including in your model. You can take any question and identify a chain of mechanisms that very quickly leads you to believe that everything in the world needs to be included in the model. Consequently, model building involves making important decisions about what components you are going to model explicitly.

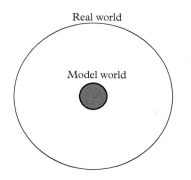

Figure 1.2 *The model world contains a subset of the elements and processes that happen in the real world. Your job is to clearly identify the boundary between the model world and the real world.*

One useful way to think about this is that you are going to create a brand-new world. You decide what is in that world, and the rules that govern the behavior of that world. As we've already noted, this world—let's call it the model world—is going to be different from the real world. In other words, the model world is a vastly simplified version of the complicated real world (figure 1.2), including only a subset of the actual processes that actually occur. You want to create a model world to shed light on the real world, and the only way to do that is to clearly delineate the difference between the two worlds. This process of delineation is what I call "defining the model world." It is only by explicitly recognizing the simplifying decisions that make it possible to distinguish between the two worlds can we possibly hope to learn anything about the real world from models.

Walters (1986) suggests that there are four kinds of decisions that you make that distinguish the model world from the real world: the model breadth, the model detail, the spatial scale, and the temporal scale. Each of these are explored below.

Breadth refers to how many of components of the real world you are explicitly modeling. And by component, I mean a naturally distinct entity, something that cannot easily be combined with another entity. For instance, climate and predation are two very distinct things. Your model might include climate, predation, climate and predation, or neither. But you certainly wouldn't have a variable called "climate & predation." Breadth is simply how many of these things (processes, drivers, etc.) that your model is going to explicitly include.

It is best understood with an example. Say you are modeling bark beetle populations in the U.S. mountain west. You might choose to include only the density of bark beetles in a particular forest stand. That density will be affected by many processes, but none of them will be explicitly represented. In that case, you are choosing to minimize the breadth of the real world by not explicitly modeling predation, forest composition and extent, environmental conditions, competing species, diseases, and so on. On the other hand, you might instead be trying to model the entire food web of this same forest stand. Here you'll track carbon as it moves among various compartments such as soil, plants, herbivores, carnivores, and so on. This model has high breadth, as you are explicitly modeling the interactions between many components of the real world, including bark beetle predators, trees, and other organisms that are affected by tree mortality. You might also include abiotic parts of the ecosystem by modeling hydrology, nutrients, and sunlight. That model would have enormously high breadth.

Detail simply means how each of the components of nature are being represented. In the bark beetle example above, you have a lot of choices for how you want to represent the bark beetle population. The model could have very low detail by assuming all bark beetles are identical, so you will only keep track of the total number of bark beetles. Alternatively, it could account for the fact that different life stages have distinct demographic rates, so each life stage is tracked separately. There could be a single bark beetle population, or you could consider several bark beetle populations connected to each other. You might consider the explicit spatial distribution of bark beetles, so that you are keeping track of population size at many different locations. You might model every bark beetle individually (very high detail), keeping track of all of its traits.

Figure 1.3 provides examples of models with emphasis on either breadth or detail. In the top panel, we're keeping track of ducks, but also explicitly modeling their predators and their food. Because we have several components of nature, this has an emphasis on breadth, but at the expense of detail. Compare this to a model with little breadth (only one component of nature represented, ducks) but more detail (the lower panel in figure 1.3). Here we've set aside any consideration of dynamic feedbacks between predators and prey so that we can model the ducks with more detail. In this model, there are different stages of ducks, each with unique traits.

Time and spatial scale round out our model delineation. These decisions are usually fairly easy once you've designed your question. For example, a model of optimal foraging decisions needs to be at the time and space scale of individual foraging activities (generally small temporal scale and short time scale), while a model of metapopulation dynamics needs to be at very large spatial and temporal scales. Although these decisions are not hard to make, it is important to be very deliberate and explicit about them.

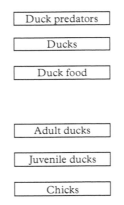

Figure 1.3 *A model with high breadth and low detail (top), and one with low breadth but high detail (bottom).*

1.3.1.2.1 Trade-offs Levins (1966) proposed three main dimensions of trade-offs that modelers must face: generality (the extent to which predictions /outcomes can be widely applied), realism (the extent to which models realistically represent the true state of the world), and precision (usually thought of as the ability of the model to predict outcomes, though, in truth, he meant, "Do you really need to know parameter values, or can you just understand the general shapes of relationships?"). It is impossible to have a model with high generality, high realism, and high precision, because the choices you make to maximize any one of these necessarily diminish the others. A model that is very generalized typically has to shed a lot of realistic detail that would be specific to any particular system. A model intended to generate precise predictions may have little realism and may only work in a limited number of systems. At best, you might get two out of these three.

Put another way, it is usually very difficult to build a model that has both high breadth and high detail. That is, if you attempt to model many different components, each modeled with a high degree of detail, the model can soon become so complex that it is no longer useful. As model complexity increases, your ability to understand why it generates specific outcomes decreases. If you can't understand why your model works, you might as well be studying the real world.

1.3.1.3 Building a conceptual model

I follow a format similar to the one described by Otto and Day (2007). This one works for me because it helps me avoid certain pitfalls, and there is no set convention in the field. For that reason, this is a good place for you to start, but I suggest focusing more on the underlying concepts rather than on the specific notation choices.

There are three main steps: (1) identify your state variables, (2) identify the rates that cause each of the state variables to change, and (3) identify which state variables affect which rates.

As an example, consider a simple model of a population where every individual is equivalent. That means we have a single state variable, the population size. Draw this with a box (figure 1.4), and label the inside of the box "N" to remind us that we'll use the notation "N" to refer to population size. Now list the rates that make population size increase: births and immigration. And do the same for the rates that make population size decrease: deaths and emigration. For each rate that causes the population to increase, put an arrow pointing into the box, and label the arrow with the name of the rate it is referring to. Do the same for rates that cause the population to decrease, but have those arrows pointing away from the state variable box (figure 1.4).

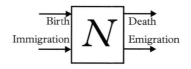

Figure 1.4 *The start of our conceptual model with one state variable and rates denoted with arrows.*

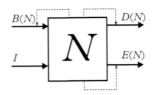

Figure 1.5 *The start of our conceptual model with one state variable, with the feedback between the state variable and the rates indicated by dotted lines.*

A central feature of any model is the relationships between rates and state variables. This is the way that different state variables interact with each other—they affect each others' rates. We need a structured approach to make sure that we don't specify that one state variable has a direct effect on another state variable. If we did this, we are essentially coding the model result directly into the model structure. For example, take the conceptual model at the top of figure 1.3. We suspect that duck predators have a negative effect on ducks, so there is a tendency is to draw a line showing that predators cause ducks to go down. But they don't affect ducks directly. They don't make ducks disappear. Rather, predators affect the duck death rate. While, in most cases, this will indeed cause the state variable, that is, the number of ducks, to decrease, there are actually some circumstances where this increase in death rate will cause ducks to become more abundant (the so-called hydra effect; Abrams 2009). The details of why and how this can happen are not terribly important. Rather, the point is that we would not even be aware of this phenomenon if we hadn't been careful in linking state variables through their effects on each others' rates.

Remember: feedbacks happen between rates and state variables, never between state variables directly.

We'll depict these feedbacks as dotted lines that link the rates to the state variables that affect them. We'll also introduce a bit of notation: B, I, E, and D, for births, immigration, emigration, and deaths, respectively. Let's assume that the number of births, deaths, and emigrants depends on population size, N, but the number of immigrants does not. This is essentially assuming that the number of individuals coming into the population from other areas is not affected by the population size. We will use the notation (N) for rates that depend on the state variable N (i.e., $B(N)$ means the number of births is some function of N). Finally, so that we remember which rates depend on which state variables, we'll draw dotted lines from each state variable to the rate that it influences (figure 1.5).

1.3.2   Advanced applications and topics

1.3.2.1   Interacting species

How do we extend the approach shown above when we have multiple, interacting state variables? Let's consider a model of the snowshoe hare (*Lepus americanus*) and the Canada lynx (*Lynx canadensis*). Hares are important prey for lynx, and lynx are an important source of death for hares. First, define the state variables: L will be the

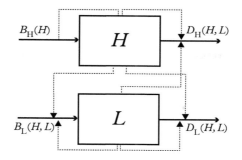

Figure 1.6 *Conceptual box-and-arrow diagram of the hare-lynx predator-prey system. Here, lynx affect hares by increasing their death rate. Hares affect lynx through both birth and deaths in this image, but often we might choose to link hares to only one of these rates.*

population size of lynx, H will be the population size of hares. We know that the same basic rates will govern the dynamics of each of them: births, deaths, immigration, and emigration. For simplicity, assume no immigration or emigration occurs, so that we only pay attention to births and deaths.

We can begin the conceptual model diagram as we did above, recognizing that births and deaths lead to increases and decreases in each state variable and that each state variable affects the birth and death process.

How do we model the effect of lynx on hares? One obvious choice is to say that the death rate of hares depends on the abundance of lynx. In other words, the total number of hare deaths will be a function of both the number of hares and the number of lynx. We might denote that as $D_H(H,L)$, which says that the number of hare deaths is a mathematical function that takes as inputs H and L (figure 1.6). Presumably, more lynx means greater death rate of hares, all else being equal.

How do we model the effect of hares on lynx? This one is less clear-cut. One line of reasoning is that if there are a lot of hares present, then lynx have plenty of prey, grow larger, store more energy, and therefore have better reproductive success. In this case, we would say that the lynx birth rate is a function of the number of lynx and the number of hares (figure 1.6), denoted $B_L(H,L)$. Alternatively, perhaps the lynx are continually on the brink of starvation. In this case, the availability of prey will reduce the death rates of lynx, so the lynx death rate is a function of lynx and hare (denoted $D_L(H,L)$. Finally, you could assume that both are true, as is done in (figure 1.6).

1.3.2.2   *What happens when two state variables share a rate?*

Consider a metapopulation, which is several populations distributed in space with movement of individuals among them. For the sake of simplicity, presume there are two of these populations, whose abundance we denote with the state variables N_1 and N_2, and that individuals from population 1 occasionally emigrate out into population 2. Modeling births and deaths might be essentially the same as what we did in section 1.3.1. But here, the number of individuals that emigrate out of population 1 is the same as the number

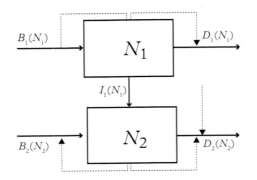

Figure 1.7 *A conceptual model for when the rate output from one state variable is a rate input to another state variable. Here, individuals migrate from population 1, N_1, to population 2, N_2, at a rate $I_1(N_1)$. This rate is "shared" as it affects the dynamic behavior of both N_1 and N_2.*

that immigrate into population 2. For that reason, we might depict that with a single arrow that connects the two state variables (figure 1.7), which reminds us that this is a "shared" rate that affects both state variables.

1.3.2.3   Translating conceptual models into math

We start by writing out general equations based on the conceptual box-and-arrow diagram. Because these are dynamic models, our models are depicting the change in the state variable over time. This means we first need to make a decision about whether we're going to model the system in discrete-time steps, or whether we're going to model the system in continuous time. For the most part, the choice depends on the underlying biology of the process being modeled, but often mathematical convenience comes into play. If you know this population has discrete generations, and reproduction is restricted to a brief period of the year, then a discrete-time model would seem to capture the biology better than a continuous model. If there are many overlapping generations, so reproduction was occurring more or less every day of the year (as would be the case for, e.g., microbes), you might choose a continuous model.

Let's revisit the population model that we started above, and assume for now that we've decided to use a discrete model. We begin by setting up an equation that denotes the change in population numbers over time on the left-hand side:

$$\frac{\Delta N}{\Delta t} = \tag{1.1}$$

We need to fill in the right-hand side by looking at our conceptual model. Whenever we have an arrow pointing into the box, we need to include that rate on the right-hand side with a positive sign. Arrows pointing away from the box depict rates that make $\frac{\Delta N}{\Delta t}$ decline, so we add those to the right-hand side with a negative sign. Following this convention and using the notation we described on the previous page, we now have

$$\frac{\Delta N}{\Delta t} = B(N) + I - D(N) - E(N) \tag{1.2}$$

Note that, so far, we haven't said anything about how we're going to represent any of the model components on the right side of the equation. That is, we said that reproduction is going to depend on population size but we haven't specified that relationship yet.

If we wanted to model this with a continuous-time model, then our equation would look like this:

$$\frac{dN}{dt} = B(N) + I - D(N) - E(N) \tag{1.3}$$

Although the right side of the equation did not change, the meanings of the rates are now different: in continuous-time models, everything happens instantaneously, so the rates change as the state variables change.

How do we go about figuring out what these functions are? This is where experience really helps out, but you can begin by drawing simple graphs that depict the general shape of each relationship. For instance, we presume that each of the rates is probably positively related to population size. In other words, the total number of deaths is probably greater when the population size is bigger than when it is smaller. Maybe we want to make reproduction a saturating function of population size, to depict some maximum number of new births into the population. Alternatively, maybe mortality rate is an accelerating function of population size, so that a greater fraction of the population dies each year when population size is big. Once we've decided on the shape of these graphs, we can begin writing the actual mathematical relationships.

1.3.2.4 *Drawing function shapes*

For each feedback in our model, we need to sketch the shape of that feedback. It is useful to distinguish between total rates (e.g., total number of births per unit time) and per capita rates (total number of births per individual per unit time). It is easy to confuse these two. Make sure that, in your model, the arrows coming out of each state variable are the total rates. For illustration, consider the relationship between the number of births and the population size. The simplest relationship would be a linear relationship (figure 1.8). If the relationship between total number of births and population size is linear, then the number of births per individual (the per capita rate) is constant (see the right-hand side of Figure 1.8).

We might instead think that total number of births will increase with population size but in a nonlinear way so that there is some asymptotic relationship (figure 1.9). Note that even though total number of births is increasing, the per capita birth rate is decreasing.

The shapes of these relationships are important, because they are implying something about ecological processes. Consider, for instance, an accelerating relationship between population size and number of births (figure 1.10). This particular function implies that the per capita birth rate is increasing with population size. This kind of relationship will

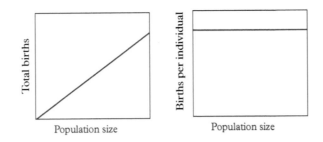

Figure 1.8 *Total birth rate as a function of population size, $B(N)$, on the left, and the implied per capita rate on the right.*

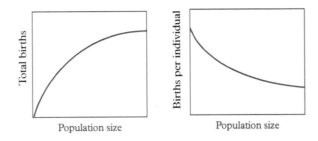

Figure 1.9 *Here total birth rate does not increase linearly with population size; this implies that per capita birth rate must decline with population size.*

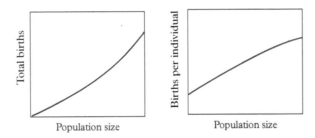

Figure 1.10 *Total birth rate is an accelerating function of population size, which implies that per capita birth rate increases with population size.*

usually not produce sensible model behavior unless per capita death rates also happen to decrease with population size.

1.3.3   The rest of the process

1.3.3.1   *Model coding and diagnostics*

This is the nuts-and-bolts part of modeling. Extensive programming skills aren't necessary for many model applications. Many questions can be explored by using spreadsheets, an approach that has the advantage of providing a visual way of checking to ensure that the numbers are matching up properly. I often evaluate components in my models in spreadsheets, especially when I'm confused about their behaviors. At the other extreme, you can code in Fortran or C++, but these require a lot of programming expertise. In-between are programming languages that are relatively easy to use, such as R, Matlab, and Python.

As you go along, it is important to develop procedures to "debug" your programs. For example, if I know that the population size can't be negative, try to push the model parameters to make sure that population size is always greater than or equal to 0. This is sometimes called model "verification," meaning you want to verify that the model is doing what it is supposed to be doing.

1.3.3.2 *Interpreting and analyzing the model's output*

Once your model is running and behaving properly, you can finally begin to use the model to answer your questions. One trap that people sometimes fall into is taking the model output as a truth of sorts and failing to think about the underlying model that generated it. This is a terrible mistake, because the learning comes from completely understanding the relationship between the model assumptions and the model predictions. Remember, in terms of the model world, you are omniscient. In some cases, you might find that your result is really just one of your assumptions. Try putting the model results and the chain of logic that produced them into plain language.

I find it useful to think of model analysis and interpretation as three separate steps: *describe, explain, and interpret.*

Describe, explain, and interpret

Describe: This is the simplest of the three, as it involves answering the question "What happened?" What was the model behavior, its predictions, its sensitivities, and so on? This is like any scientific enterprise. Before interpreting the results of your experiment, you would likely first describe what happened in the experiment. In modeling applications, we can easily become overwhelmed by the number of things that we might want to describe, but having a well-defined purpose helps guide you in choosing what model results to focus on. At this point, you are still in the "model world" because you are describing properties of the model, not the real world.

Explain: How did the internal logic of the model gave rise to the patterns you described? Put another way, why did the model do what it did, given how you defined the model world? The great thing about models is that you have perfect knowledge of the model world, so you should be able to explain why everything happened.

continued

Interpret: After you've described the model behavior, and you understand why the model generates the observed behavior, you can use that knowledge and apply it to the real world. One way to approach questions that ask, "What does this mean with respect to the real world?" is to do the following: how does your knowledge of the model world shape the way you view the real world? Put another way, what new insight do you have from exploring the model that helps you understand what might be happening in the real world?

Sometimes these interpretations will be very literal. "If the real-world process described by parameter X is large, then we will get real world behavior like this." More commonly, we're looking for something less literal. Given what you know about how this simple model system works, what can you say about the complicated real world?

This process of *Describe, Explain, and Interpret* is not easy and takes practice. One of the main goals of this book is to help you develop these skills.

1.3.4   Modeling is not a linear process

You will likely find that as you describe, explain, and interpret, you discover that there are parts of the model that you need to change because they lead to bizarre or unrealistic predictions. This is an important part of the learning process and is a valuable part of modeling. It is not uncommon to have a series of seemingly reasonable assumptions that, when put together, produce unreasonable conclusions. This is cause for you to rethink your assumptions and identify holes in your logic that would not be apparent in a simple verbal model. You may even find that you need to revise your question!

Summary

- Models are purposeful simplifications of reality.
- Models need to be faithful to reality, but "realism" alone is not a useful way to judge a model.
- A good model is one that serves its purpose well.
- Dynamic models start by identifying the breadth, detail, and spatial scale of the model world that distinguish it from the real world.
- To avoid circular reasoning, follow a process whereby state variables affect rate variables, and rate variables, in turn, affect state variables.

Exercises

1. Lynx and coyotes compete for shared prey (snowshoe hare) in Canada and Alaska. You want to create a model to explore the relative strengths of intra- and interspecific competition that allow these two species to coexist. For your answers below, consider the simplest model possible that might be useful for addressing this question. Bound the problem by describing the minimum breadth (extent) and detail of the model needed to tackle this question. Identify your state variables and rates. Create a box-and-arrow conceptual diagram of the model. Use your box-and-arrow diagram to describe the mathematical dynamic equations for the biological component of the model only. Draw graphs that depict the general shapes of the relationships between state variables and the rates in your model.

2. You are interested in modeling the spread of disease because you want to know the best way to reduce the numbers of deaths that the disease causes. There are two ways to do this: you might invest in reducing the transmission of the disease, or you could invest in treating infected individuals to reduce the death rate of those infected. Presume that you are interested in implementing policy measures that operate over fairly short time horizons, so you are not interested in tracking reproduction or deaths other than those caused by this disease. You know that individuals who have not yet had the disease are susceptible to infection. Denote the number of susceptible individuals as N_s. Susceptible individuals may contract the disease and become infected individuals. Denote the number of infected individuals as N_i. Infected individuals have two possible fates: either they recover from the disease and are forever immune, or they do not recover and subsequently die. Denote the number of individuals that survive and are immune as N_r, and the number of individuals that have died as N_d. The rate at which susceptible individuals become infected (infection rate, I) depends on both the number of susceptible individuals and the number of infected individuals, at any given time. The rate at which infected individuals recover from the disease (recovery rate, R) depends on the number of infected individuals. The rate at which infected individuals die from the disease (death rate, D) depends on the number of infected individuals. Based on this, would you use a discrete-time model or a continuous-time model? Explain. Identify your state variables and rates. Create a box-and-arrow conceptual diagram of the model. Interpret your conceptual diagram in the form of mathematical dynamic equations. Draw graphs that depict the general shapes of the relationships between state variables and the rates in your model.

3. Mussels and barnacles compete for space in intertidal regions. Explain in one or two sentences how you could model mussel and barnacle competition without explicitly modeling "space" as a state variable.

2

Introduction to Population Models

This chapter, combined with the next, illustrates how one can take one level of breadth (here, a single population) and explore it across a range of detail. You might find the diversity of model assumptions, functions, and structures to be a bit overwhelming. "How do I know which to choose?" If you feel this way, remind yourself that the goal here is to show you how modeling a single entity, population density, can be done in many different ways, depending on the purpose of the model. This is building up your modeling toolbox so you can later make good modeling decisions.

2.1 Introduction to population dynamics

Understanding the dynamics of populations remains one of the fundamental goals of ecology. Not surprisingly, many models have contributed to the theory of population dynamics and regulation. They vary considerably in terms of model depth, breadth, intended uses (e.g., prediction vs. generality), and structure. This chapter will largely focus on the behavior of simple models, to see how intrinsic factors can dictate variability in population size.

In 1949, Varley published a now-famous data set showing the dynamics of four insect species (Varley 1949; see figure 2.1). *Note the logarithmic scale on the y-axis.* As you can see, these populations undergo dramatic fluctuations, with population size varying by two to three orders of magnitude within short (five-year) periods. This raises some important questions:

- What governs the average population density?
- What is the relative importance of intrinsic factors (internal to the population) versus extrinsic factors (external, such as environmental conditions) factors in driving variability in population densities?
- How predictable are population densities from one year to the next?
- What kinds of dynamic behavior are possible from simple ecological processes?

Introduction to Quantitative Ecology: Mathematical and Statistical Modelling for Beginners. Timothy E. Essington, Oxford University Press. © Timothy E. Essington 2021.
DOI: 10.1093/oso/9780192843470.003.0002

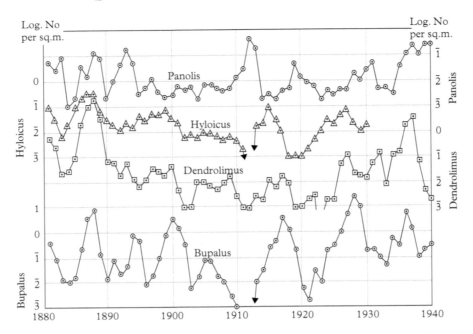

Figure 2.1 *Population fluctuations of four insect species deemed to be "pests" in German forests. From Varley 1949, Journal of Animal Ecology, Vol 18. British Ecological Society, John Wiley & Sons Ltd. © 2020 The Authors. Journal of Ecology published by John Wiley & Sons Ltd on behalf of British Ecological Society.*

In this chapter, we'll focus on a subset of these questions. We'll explore the process of developing a population model and examine what kinds of dynamics are possible from population models and how this depends on density dependence in demographic rates. Later (in chapter 3), we'll learn how to develop models that contain more biological detail, by accounting for variation in demographic rates by age or life-history stage.

2.2   Fundamental structure of population models

To begin, we'll develop models where all individuals within a population are identical to each other. We began developing such a model in the previous chapter, with a box-and-arrow diagram (figure 1.5).

Assume for now that immigration (I) and emigration (E) both equal 0. To turn a conceptual diagram (figure 1.5) into a mathematical equation, we'll assume that the model works in discrete-time steps (more on this later):

$$\frac{\Delta N}{\Delta t} = B(N) - D(N) \tag{2.1}$$

This equation simply says that the change in population size from one time step to the next is the difference between the number of births (B) and the number of deaths (D). It also says that the number of births and deaths both depend on the population size. Most of the alternative models that you'll see are simply different ways of describing what these functions might be.

We should address one last notational issue before we move on. For discrete-time models, we usually want a model that tells us the population size at one point in time as a function of population size at an earlier time (typically the previous time step):

$$N_{t+t} = N_t + \frac{\Delta N}{\Delta t} \tag{2.2}$$

where we will substitute in some function for $\frac{\Delta N}{\Delta t}$. We'll refer to this as the *recursive equation*. It serves as the backbone for all of our discrete-time population models. It is also often called a "difference" equation, to distinguish it from a differential equation. I find that notation somewhat misleading, because, in my mind, the discrete-time version of the difference equation is similar to equation 2.1. The important thing to know is that the recursive equation propagates forward the population size from time t to time $t+1$, where 1 could be whatever discrete-time interval makes sense for your model.

2.3 Density-independent models

The rates described above, $B(N)$ and $D(N)$, are the total number of births and deaths that occur over some time interval. We often think about rates on a per capita basis. Per capita literally means "per head," so per capita rates describe the number of births per individual, and an individual's probability of death. When per capita rates are constant regardless of population size, we say that these rates are *density independent*. In other words, the per capita rates are independent of population density. By extension, if *all* per capita rates in our model are density independent, we say that our model is density independent.

If we denote b and d as the per capita birth and death rates, respectively, then the functions describing the total number of births and total number of deaths are, respectively,

$$B(N) = bN \tag{2.3}$$
$$D(N) = dN \tag{2.4}$$

We can then substitute these functions into our dynamic equation:

$$\frac{\Delta N}{\Delta t} = bN - dN = (b-d)N \tag{2.5}$$

Finally, we need to turn this into a recursive equation that will calculate N_{t+1} based on N_t. In a discrete-time model, this is fairly simple. Presume that the Δt in our dynamic equation corresponds to one time increment (e.g., one year). The population size in the next time step, N_{t+1}, equals the population size in the previous time step, N_t, plus the increment of change, which is given by our dynamic equation:

$$N_{t+1} = N_t + \frac{\Delta N_t}{\Delta t} = N_t + (b-d)N_t \tag{2.6}$$

We can simplify this a little bit:

$$N_{t+1} = (1+b-d)N_t \tag{2.7}$$

For even more simplicity, we can define a new parameter, λ, and set it equal to $1+b-d$. Given this new parameter definition, we can rewrite the recursive equation as

$$N_{t+1} = \lambda N_t \tag{2.8}$$

The parameter λ is the *population growth rate*. We can use this model to explore the types of population dynamics that are possible in the absence of random events or density dependence. First, let's presume that the per capita birth and death rates are equivalent ($b = d$). When that's true, then $\lambda = 1$, so that the recursive equation simplifies to $N_{t+1} = N_t$. This just says that the population stays at the same level for every time step. This trivial example also reminds us that when the birth rate equals the death rate, the population will not change (this will be helpful for estimating steady-state population sizes in the density-dependent models that follow).

Reparameterization

In the density-independent model, we derived an expression by thinking about birth and death rates, and then replaced that expression with a simpler one that uses different parameters, here λ in place of $1+b-d$. We will do this repeatedly in this book: we will take a model, introduce one or more new parameters, and then "reparameterize" our models.

We do this to draw emphasis to the parts of the model that are important. But we are not changing the underlying model at all.

As an analogy, consider the following two sentences:

A woman was chased by a dog.
A dog chased a woman.

Both convey the same information: there was a woman, and a dog, and a chase. But the subject of each sentence is different, placing emphasis on different parts of the scene. The first sentence is focusing on the woman, while the second is focusing on the dog.

Reparameterization works in the same way: the information has not changed, but the emphasis has.

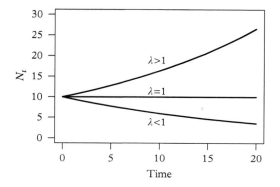

Figure 2.2 *The three possible behaviors of a density-independent model.*

An important thing to realize about density-independent models is that the range of dynamic behavior they can produce is pretty limited. In fact, there are only three different types of behavior: no change (when $\lambda = 1$), exponential decline (when $\lambda < 1$), and exponential increase (when $\lambda > 1$) (figure 2.2).

Notation

How we denote time in our models is among the most important notation issues. Generally, with discrete-time models, one refers to the value of a state variable N at time t by using the notation N_t (if you were to read it out loud, you might say "N at time t").

For continuous-time models, one often denotes state variables as $N(t)$ to indicate that our state variable is a function of t. If you were to read it out loud, you would say "N of t."

I will occasionally drop the time notation in differential and difference models when the time is clearly implied. For instance, rather than:

$$\frac{\Delta N_t}{\Delta t} = f(N_t)$$

I might instead write

$$\frac{\Delta N}{\Delta t} = f(N)$$

While this might be confusing if the difference equation depends on population sizes in prior time steps, we'll not cover any of models that do so in this book. Many find the simpler notation clearer and easier to follow.

Given that the behavior of density-independent models is so limited, why do we use them at all? One reason is that we use them as the backbone of more complex models. Later (in chapter 3), we'll add more detail to these models through population structure. Even later (in chapter 5), we'll add complexity to these density-independent models to understand how variation in demographic rates affects populations. Also, we use these models in cases when we believe populations are so low that density dependence is probably not a major influence. Take, for instance, the short-tailed albatross. For years, they were hunted nearly to extinction. The one Pacific island where they escaped extinction just happens to be an active volcano. But because of protection efforts, their population has grown largely at a constant, density-independent rate from 1947 to the early 2000s. Despite their simplicity, these density-independent models are frequently useful.

Applying the describe, explain and interpret framework to the density-independent model

Describe: The density-independent model has three possible behaviors: exponentially increase, exponentially decrease, and stay exactly the same

Explain: This model has no feedback between state variable and the per capita growth rate. As a result, this rate remains the same even when population grows to high levels or crashes to low levels. When births perfectly match deaths, there is no change in the population.

Interpret: In the absence of any feedbacks between rates and state variables, model behavior is limited and is likely unrealistic. This suggests that most real-world populations likely have some feedbacks between population size and demographic rates.

2.3.1   Continuous-time density-independent models

We approach the development of density-independent models the same way if we wish to assume a continuous-time model. Again, assuming no immigration or emigration, and b and d represent birth rates, then

$$\frac{dN(t)}{dt} = bN(t) - dN(t) \tag{2.9}$$

$$= rN(t) \tag{2.10}$$

where r is the difference between per capita birth and deaths—it is expressed in units of inverse time (time^{-1}). Note here the subtle difference in the interpretation of b and d in this model and in equation 2.8. There, d was the fraction of individuals that die over a defined, discrete-time interval, say, one year. It must necessarily be between 0 and 1. Here, d can be any number greater than 0. It can be 0.1, 1, or even 10 for species with extremely high mortality. How can this be? Because the model is continuous, so change happens at every instant (for this reason, you might see these referred to as "instantaneous rates").

Solving for *N(t)*

I promised I wouldn't make you do an integral, but it wouldn't hurt to see an application of an integral. Given the differential equation above, we can rearrange

$$\frac{dN}{rN(t)} = dt$$

and then take the integral of each side:

$$\int \frac{dN}{rN(t)} = \int dt$$

This gives us

$$\frac{\log(N(t))}{r} = t + C \tag{2.11}$$

where C is the integration constant. We solve for C by defining $N(0)$ as the value of $N(t)$ when $t = 0$. We then rearrange to get

$$N(t) = N(0)e^{rt}$$

This model has a nice, tractable mathematical solution (as explained in "Solving for $N(t)$"), and an intuitive interpretation of the parameter r. When $r > 0$, the population grows; when $r = 0$, the population stays the same; and when $r < 0$, the population shrinks.

2.4 Developing a density-dependent model

The density-independent model produced limited types of behavior, which tells us that we probably need to make either the per capita death rates or the per capita mortality rates some function of population density. For example, the per capita birth rate could depend on population density in some way: $B(N) = b(N)N$. Usually, we presume that per capita birth rates are high when populations are at low abundance and that they decline when populations are at high abundance.

The per capita death rate could also depend on density: $D(N) = d(N)$. In this case, we would want $d(N)$ to be an increasing function of population size, to capture the fact that death rates due to diseases or to food or space limitation are greater at high abundance low abundance.

We don't really know whether to put density dependence in the birth rate term or the mortality term. For the model explorations that we'll be doing, it doesn't really matter, so we instead consider density dependence in the difference between per capita death and birth rates:

$$\frac{\Delta N}{\Delta t} = [b(N) - d(N)]\,N = f(N)N \tag{2.12}$$

where $f(N)$ is some function of N that describes the difference between the birth rate and the death rate. Many different choices of $f(N)$ are possible, and we can explore these different functions to see how the population dynamics depends on the choice of function $f(N)$ and on the values of the parameters that make up $f(N)$.

2.4.1 The logistic model

Perhaps the most familiar density-dependent model, the logistic model, presumes that the difference between the birth rate and the death rate declines linearly with population size (N):

$$f(N) = r - \beta N \tag{2.13}$$

Here, the maximum per capita difference between the birth rate and the death rate is r, and this difference declines with increasing population size via a slope equal to β (figure 2.3). We can adjust the notation slightly to make the parameters more biologically meaningful. That is, if I were to tell you that my favorite population of marmots has a

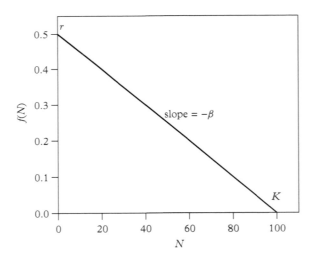

Figure 2.3 *The logistic model assumes that per capita productivity declines with population size as a straight line.*

β equal to 0.0032 marmot^{-1} year^{-1}, that probably wouldn't give you much meaningful insight about this population. Instead, we can determine the population size at which $f(N)$ equals 0 (when there is no population growth or decline). We'll call this population density K, to represent carrying capacity. We solve for this by replacing N with K above, setting $f(K)$ equal to 0, and solving for K.

Take a moment to work through this algebra. You should find that $K = r/\beta$.

We then substitute r/K for β to get our more useful version of the model:

$$f(N) = r - \beta N = r - \frac{rN}{K} = r\left[1 - \frac{N}{K}\right] \tag{2.14}$$

Now let's finish this off by inserting our expression in for $f(N)$:

$$\frac{\Delta N}{\Delta t} = rN\left[1 - \frac{N}{K}\right] \tag{2.15}$$

And we'll plot this expression as a function of population abundance and the resulting recursive equation (figure 2.4).

The following are important things to remember about the logistic model:

– It does not distinguish whether density dependence occurs in births or deaths (though it can be derived by assuming either b or d is a linear function of density as long as the other one is constant).

– It assumes a linear decline in the difference between per capita birth and death rates.

– The parameter K is derived by simple reparameterization, to make the model parameters more meaningful.

Learn how to code the logistic model in Excel (section 12.1 [pp. 188–90]) and in R (section 13.2, [pp. 207–9].)

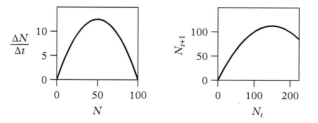

Figure 2.4 *Rate of population change and the recursive equation are domed-shaped functions of population abundance in the logistic model.*

2.4.2   Advanced: Continuous logistic models

As we did in section 2.3.1, we can develop a continuous-time version of the logistic model. We start with the same assumptions, where the parameter r is the intrinsic growth rate, the maximum difference between birth and death rates, and K is still the carrying capacity:

$$\frac{dN(t)}{dt} = rN(t)\left(1 - \frac{N(t)}{K}\right) \tag{2.16}$$

To solve for $N(t)$, we have to do some calculus, so that if you did the considerably more difficult integration, you would get the following solution:

$$N(t) = \frac{N(0)Ke^{rt}}{K + N(0)(e^{rt} - 1)} \tag{2.17}$$

The equation gives you the exact population size for any time t, given an initial population size $N(0)$.

Learn how to numerically simulate any set of ordinary differential equations in section 14.5 (pp. 243–49).

2.4.3   Other density-dependent models

There are a multitude of functions $f(N)$ that one might use to generate density dependence. Below are some common ones. You might rightly ask, "How do I know which one to use?" Please have some patience (as we will explore this question much later in the book). For now, we are seeing what kinds of model functions have been used and how the behavior of models depends on the shape of these functions. That is, we are not presently concerned with "which model function is correct?" but rather we want to know "what are the consequences of choosing different functions?"

2.4.3.1   The Ricker model

Under the Ricker model, the per capita death rate is often presumed to equal 1, and per capita birth rates are a log-linear function of N: $\alpha e^{-\beta N}$. Here $\alpha - 1$ is equivalent to r in the logistic model. Under these assumptions, the recursive equation is

$$N_{t+1} = N_t \alpha e^{-\beta N_t} \tag{2.18}$$

This model is often used for semelparous species (species who reproduce only once).

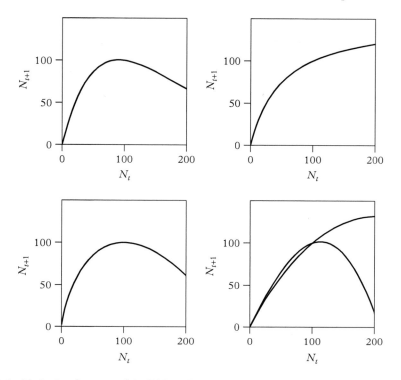

Figure 2.5 *Clockwise from top left: Ricker, Beverton-Holt, theta-logistic, and Gompertz models, expressed as recursive equations. All have the same equilibrium at 100. Two values of θ (0.75 and 1.5) are shown for the theta-logistic model.*

The model can be slightly reparameterized with respect to carrying capacity, K, which is the level of N for which $N_{t+1} = N_t$. This turns into a useful expression:

$$N_{t+1} = N_t e^{\alpha_l \left(1 - \frac{N_t}{K}\right)}$$

where α_l is equal to $\log(\alpha)$ (figure 2.5).

2.4.3.2 The Beverton-Holt model

In the Beverton-Holt model, the per capita birth rate is a hyperbolic function of population density, and the death rate can take any plausible value between 0 and 1:

$$f(N) = \frac{\alpha}{1 + \beta N} - d$$

$$N_{t+1} = N_t + N_t \left(\frac{\alpha}{1 + \beta N} - d\right) \tag{2.19}$$

2.4.3.3   *The Gompertz model*

The Gompertz model is based on a recursive equation in log-space, which can be expressed in two different ways:

$$\log(N_{t+1}) = \log(N_t) + b\log\left(\frac{N_t}{K}\right) \tag{2.20}$$

$$N_{t+1} = N_t\left(\frac{N}{K}\right)^b \tag{2.21}$$

2.4.3.4   *The theta-logistic model*

In the theta-logistic model, there is a slight addition to the logistic function. The function $f(N)$ is modified by taking the term N/K and raising it to some power, θ:

$$f(N) = r\left(1 - \left(\frac{N}{K}\right)^{\theta}\right) \tag{2.22}$$

$$N_{t+1} = N_t + rN_t\left(1 - \left(\frac{N_t}{K}\right)^{\theta}\right) \tag{2.23}$$

The theta-logistic model can have more diverse shapes than the logistic model can. Namely, the product $Nf(N)$ will peak at $0.5K$ when $\theta = 1$, will be less than $0.5K$ when $\theta < 1$, and will be greater than $0.5K$ when $\theta > 1$ (see figure 2.5).

2.4.4   Allee effects

All of the above density-dependent models are "compensatory," meaning that $f(N)$ is always a declining function of N. We call them compensatory because the feedback implies that as N increases, $f(N)$ compensates for the changes in abundance.

But models do not need to be compensatory for all values of N. We can instead consider a version of $f(N)$ where $f(N)$ is an increasing function of N for a range of small population sizes. In other words, when N is very low, increasing N leads to increased per capita population growth. This kind of relationship is often called an "Allee effect," named after Warder Clyde Allee, who found that multiple animals benefited from forming high-density aggregations (Allee 1928). In other fields, it is called "depensation" (Liermann and Hilborn 1997).

Graphically, the distinction between compensatory and Allee (depensatory) functions $f(N)$ can be seen in figure 2.6. Here the compensatory function is simply the logistic function, where the per capita population growth rate declines linearly with population size, crossing the x-axis at $N = 1000$ (this is the carrying capacity, K). The Allee function has the same carrying capacity and also increases with decreasing population size from about K to about $N = 600$. At that point, the per capita production begins to decline and crosses 0 at $N = 200$. Below this point, population growth is negative. We'll call this threshold point A, for the "Allee threshold."

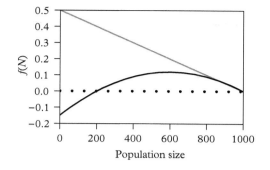

Figure 2.6 *Compensatory feedback (gray) versus Allee (black) feedback.*

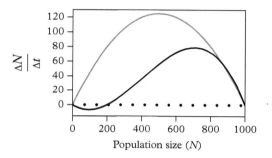

Figure 2.7 *The Allee model (black line) predicts very different relationship between population abundance and population rate of change than the logistic model (gray line) does.*

Lewis and Karieva (1993) provided a clever way to model a population with Allee effects.

$$f(N) = r\left(1 - \frac{N}{K}\right)\left(\frac{N}{K} - \frac{A}{K}\right) \tag{2.24}$$

It's helpful to plot the product of $f(N)$ times N to see what the resulting $\frac{\Delta N}{\Delta t}$ will be for all N_t (figure 2.7). This model predicts behavior that is very similar to that predicted by the logistic model when N is close to equilibrium (here, greater than about 800) but predicts reduced population growth for smaller values of N, and negative growth when N is less than 200.

2.5 Dynamic behavior of density-dependent models

A key point of this section is that very simple density-dependent models, such as the logistic model that we just derived, can produce very complex population trajectories. No doubt you have seen, somewhere in your previous ecological training, a population

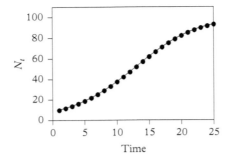

Figure 2.8 *A typical way to illustrate logistic population dynamics.*

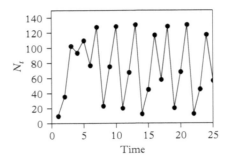

Figure 2.9 *Very complex dynamics produced by the logistic model.*

time trajectory (figure 2.8) attributed to the logistic model (here $K = 100$, $r = 0.2$, and the initial population size is 10).

This depiction of the logistic model is so common, you might think that was the only behavior it can produce. Yet, this same model, with different parameter values, can produce very complicated population trajectories (figure 2.9).

Next, we'll learn a technique called "cobwebbing" that will allow you to visualize the internal workings of the model to understand how it produces its dynamic behavior.

2.6   Cobwebbing

Cobwebbing is a graphical way of visualizing both the model dynamic behavior and the underlying model structure that gives rise to this behavior. We use it to better understand how models produce complex or simple behaviors. It's called cobwebbing for reasons that will soon be apparent.

We start with our recursive equation for the logistic model:

$$N_{t+1} = N_t + r N_t \left[1 - \frac{N_t}{K} \right]$$

We then make a plot of this recursive equation, with N_{t+1} on the y-axis, and N_t on the x-axis. Now we add a line that shows where $N_{t+1} = N_t$ (figure 2.10). We often call this line the "replacement line," because it shows us how many individuals in time $t + 1$ are needed to "replace" individuals in time t.

To start the cobwebbing, we begin with population size in the first time step. We'll assume $N_1 = 20$. We want to know what N_2 is, given an initial size of 20. Visually, we can find $N = 20$ on the x-axis and then find the corresponding spot on the gray replacement line (figure 2.10). Because the plot shows us N_{t+1} for any value of N_t, we can simply draw a vertical line upward from the 1:1 line to the black curve that depicts the logistic recursive function. The intersection of the vertical line with the logistic function gives us population size in the second time step, N_2.

Next, we want to know what the population size will be in time step 3 (N_3). This is easily done by repeating these steps once again. So, from N_2, we move vertically from

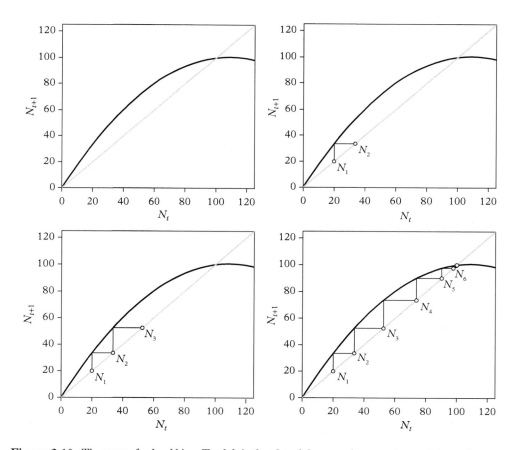

Figure 2.10 *The steps of cobwebbing. Top left is the plot of the recursive equation and the replacement line. Top right shows one step of cobwebbing, bottom left shows the first two steps, and bottom right shows all cobwebbing steps to equilibrium.*

the replacement line up to the logistic curve and then move horizontally over to the replacement line to get N_3 (figure 2.10).

And we repeat this for as many time steps as we want. In figure 2.10 I've labeled all of the N_t so you can see the changes in population size. At first the N_t are pretty close together (slow population growth), then they become spaced far apart (the phase of rapid population growth), and as the population starts to approach the carrying capacity ($K = 100$), they get closer together again.

Learn how to program cobweb diagrams in section 14.1.2 (pp. 221–23).

2.6.1   Interpreting cobweb diagrams

Figure 2.11 shows a pair of graphs with the cobweb diagram on the left, and the usual plot of population density versus time on the right. This pair is obtained by setting $r = 0.25$. Here, the typical logistic model dynamic is produced, which we can understand by seeing that the rates of population growth are modest and slowly decline as the population smoothly increases in abundance to the carrying capacity.

Now let's increase r to something quite high: $r = 1.75$ (figure 2.12) . Now our population trajectory has more complexity. The population increases rapidly, then alternatively increases and decreases in each year until it finally settles down on equilibrium. The reason for this is fairly easy to see in the cobweb diagram: the high rate of population growth leads to the population overshooting the carrying capacity, causing the population to decline for one time step. But when it declines, it undershoots the carrying capacity. And then it overshoots the carrying capacity again, each time getting closer to the actual carrying capacity. We call these "dampened oscillations."

We can increase the population growth rate even further, making $r = 2.25$ (figure 2.13). Now, the population doesn't show any tendency to settle down at the carrying capacity. Instead, it keeps overshooting and undershooting. We can see the

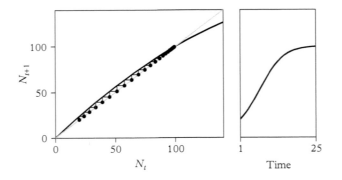

Figure 2.11 *Cobweb diagram of a logistic model when the model smoothly approaches the carrying capacity.*

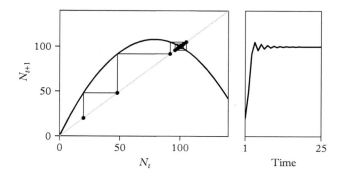

Figure 2.12 *Dampened oscillations produced when r equals 1.75.*

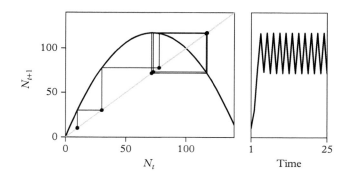

Figure 2.13 *Full cobweb diagram from the logistic model when r = 2.25.*

cobweb diagram a bit more easily if we trim off the early time steps while the model is still working things out and only show the time steps where it has settled into a pattern (figure 2.14).

So after several generations, this model predicts a population that switches back and forth from high abundance to low abundance, for the same "overshooting" and "undershooting" reason that we saw above. The only difference is that now these overshoots and undershoots persist, that is, they don't dampen through time. We call this type of behavior a "limit cycle." In this case, we have a two-point limit cycle, because the cycle consists of two alternating points.

Let's keep going and increase *r* some more. Figure 2.15 shows *r* = 2.45 (with the initial years again removed from the cobweb plot). Now our limit cycle is a bit more complex. Instead of having two points, it has four points: a super-high point, a high point, a low point, and a super-low point. This is a "four-point limit cycle."

It turns out we can keep increasing *r* to get eight-point, sixteen-point, and so on, limit cycles. But, pretty quickly, we get the result shown in figure 2.16. The plot on the right is now very different. What happened? Well, instead of a two-point cycle or a

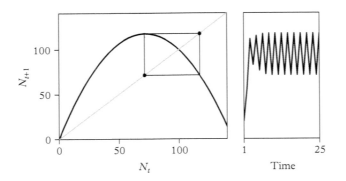

Figure 2.14 *The same model as in figure 2.13, but with the initial time steps removed so that the limit cycles can be seen more clearly.*

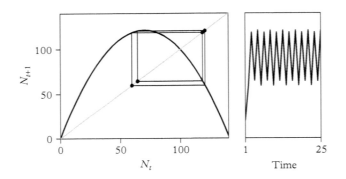

Figure 2.15 *A four-point limit cycle, illustrated via a cobweb plot.*

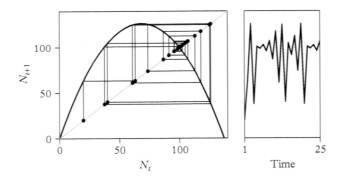

Figure 2.16 *Deterministic chaos produced by the logistic model when r > 2.75.*

four-point cycle, we have a pattern that never repeats itself: in a manner of speaking, it is a cycle of infinite length.

This is called "deterministic chaos." The two key things to remember about deterministic chaotic dynamics is that each point N_{t+1} is perfectly determined by N_t (though the population time series looks to be random), and the trajectory depends on entirely on the initial conditions. In other words, if we started with a slightly different N_t, we would get a completely different trajectory.

So, what do we make of this? Do we conclude that every complex population dynamics time series, like those shown by Varley, arises through chaotic dynamics and that we should expect populations to undergo distinct population cycles? No, that interpretation would be too literal. Rather, this exercise reveals a surprising truth (May 1976). Density-dependent feedbacks are often viewed as a stabilizing force, something that helps dampen the effect of environmental noise on populations. That is, a bad year that causes the population to decline will induce an increase in per capita productivity, thereby allowing the population to return to its previous state. The surprise is that this same process—density-dependent feedback—can also *cause* population fluctuations. May's exploration of these model behaviors therefore provided a richer understanding of population regulation and population dynamics than our simple intuition could provide.

Describe, explain, and interpret: The dynamic behavior of the logistic model

Describe: When the logistic function is steeply sloping downward when N is near K, the model both over- and undershoots the carrying capacity. This leads to limit cycles of varying length, gradually leading to chaos.

Explain: The density dependence and the population growth rate interact to produce alternatively too many individuals (above K), or too few individuals (below K). Essentially, the population growth rate is overly sensitive to population density, so it responds strongly to changes in population density. This produces unexpected swings in abundance

Interpret: Normally, when we see fluctuations in population abundance, we think that there must be some outside force at play. It is possible that the fluctuations may be due, in part, to density-dependent feedbacks within the population. Also, a population that is otherwise stable might become less so if its population growth rate is enhanced.

2.6.2 Advanced: What governs dynamic behavior?

We saw very convincingly that changing model parameters can produce surprising changes in model behaviors. There are three major kinds of behavior:

- smooth (monotonic) approach to equilibrium
- dampened oscillations that eventually reach equilibrium
- persistent limit cycles or deterministic chaos that never reach equilibrium

Is there any way to know when each of these will happen, based on the model parameters? If you want to skip very far ahead, you can read section 4.4.3 for deeper mathematical theory. Here we'll explore the question in a more qualitative way and then draw linkages later.

For all of the population models, the type of dynamic behavior depends on the slope of the recursive function at equilibrium. That is, we have an expression for N_{t+1} versus N_t. If we take the derivative of the function with respect to N and evaluate the value of the derivative when N equals K (or whatever the equilibrium population abundance is), we will know which of the three behaviors will happen.

We start with the logistic model. The derivative is found as follows:

$$N_{t+1} = N_t + rN_t \left(1 - \frac{N_t}{K} \right)$$
$$\frac{dN_{t+1}}{dN_t} = 1 + r - \frac{2rN}{K}$$
$$= 1 - r \tag{2.25}$$

The last line shows the result when we substitute K for N.

Let's think a little bit about what the derivative means. It tells us how the value of N_{t+1} changes when N_t is changed a little bit. So, consider the case when N_t equals K, but we add a very small number to N. Let's call this little extra bit n_t. We want to know whether the n_t smoothly become smaller (i.e., the population smoothly approaches equilibrium), whether the n_t becomes smaller but through dampened oscillations, or whether the n_t never becomes smaller but instead turns into the limit cycles.

Later (section 4.4.3), we'll see that we can predict how the n_t's will change through time by applying something called a Taylor approximation. But, for now, it is fine to know that the n_t will change through time approximately as

$$n_{t+1} = sn_t \tag{2.26}$$

where s is the slope of the recursive equation when $N = K$.

If s is between 0 and 1, n_t will get smaller each time step—always. So, any model parameterization that makes $0 < s < 1$ will result in a smooth approach to equilibrium. In the logistic model, $s = 1 - r$, which means that means any value of r less than 1 (with the biologically reasonable constraint that $r < 0$) will produce smooth approaches to equilibrium.

What happens when s is less than 0 but greater than -1? For the purpose of illustration, let's say s is -0.5, and the value of n_t equals 10. That is, the population size is 10 greater than equilibrium. What will the population size be in the next time period?

Well, $n_{t+1} = -0.5$ times 10, which equals -5. This means that the population size is 5 less than equilibrium. And what about the next time step? It will be -0.5 times -5, which is 2.5, meaning the population size is 2.5 above equilibrium. This will continue; each time, the population size will alternate above and below equilibrium, but the absolute value (the magnitude) of the deviation will shrink. In other words, the model will have dampened oscillations.

Finally, what happens when s is less than -1? If we take our example above but make s equal to -1.25, $n_{t+1} = -1.25$ times 10, which equals -12.5. Not only did we switch the sign of n_t, but we also moved further away from equilibrium! This will continue for a while, until our approximation in equation 2.26 no longer holds, and the model will either settle in on a limit cycle of some kind or never settle and enter deterministic chaos.

This may sound a bit esoteric, but this same reasoning will apply to any density-dependent model: Ricker, Beverton-Holt, Gompertz, theta logistic, insert-your-favorite-model-name-here. The slope of the recursive equation at equilibrium will tell you which model behavior you will see.

Summary

- All population models include the same basic processes.
- Density-independent models can only have three types of behavior.
- Density dependence can be modeled in many ways, as long as there is feedback between rates and population density.
- Density-dependent models can exhibit complex dynamics through very simple feedbacks.

Exercises

1. Get familiar with population modeling by using Excel or R to project for an unstructured, density-independent population for 50 years.
2. Use Excel or R to code a model that will run projections of a population for 100 years, using the Ricker population model:

$$N_{t+1} = N_t e^{\alpha_l\left(1 - \frac{N_t}{K}\right)}$$

continued

Examine the model behavior for a range of values of α, Apply the describe, explain, and interpret scheme to your model, particularly comparing this model behavior to the logistic model's behavior. Your interpretation should say something about the generality of the findings that is likely to hold true regardless of which model function you choose.

3. The continuous-time logistic model will not undergo cycles or deterministic chaos. Why do you think that is the case? If unsure, try programming the model by using equation 2.17 and make r a large value (>2.75, where the discrete-time model exhibited deterministic chaos). (Note: refer to the skills section in section 14.5.)

3

Structured Population Models

The models in chapters 1 and 2 presumed that all individuals were identical to each other. In many populations, several generations may be present simultaneously, and there may be important differences among individuals of different ages. Age- or stage-structured models explicitly account for these differences.

In the language we introduced earlier, structured population models have more detail than their nonstructured counterparts. They account for the differences among individuals within a population, usually by explicitly modeling them as distinct state variables. So, our single population may now be represented with two, three, or maybe 20 state variables that each depict the numbers of individuals in each population group.

As we'll see, this additional complexity means we often look to simplify other parts of the model. Much of the following material examines structured models that are density independent. By eliminating density dependence, we can more tractably handle the additional mathematical complexity that the additional detail requires.

3.1 Types of population structure: Age versus stage structure

A population is said to be "stage structured" if it undergoes distinct shifts in habitat use, morphology, behavior, or any aspect of its ecology throughout its life. An obvious example is a butterfly. It undergoes three stages: larvae (caterpillar), pupa, and adult. Werner and Caswell (1977) illustrates more complex structure, using teasel, a pretty yet prickly plant endemic to Europe but now invasive in North America and elsewhere. Teasel can be modeled in eight stages; three different stages for seeds, four different nonflowering stages, and flowering plants (Werner and Caswell 1977). Of course, other populations may have more subtle structures. Different-sized animals may have very different fecundities or survivorships, for example (Ellner and Rees 2006).

"Age structure" is a special case of stage structure, where the different "stages" are different ages. In other words, an age-structured model keeps track of the number of individuals in each age.

When would you use one over another? One useful analogy is the difference between a child's age and her grade. From one year to the next, the child will either be one year

Introduction to Quantitative Ecology: Mathematical and Statistical Modelling for Beginners. Timothy E. Essington, Oxford University Press. © Timothy E. Essington 2021.
DOI: 10.1093/oso/9780192843470.003.0003

older or not alive. This is age structure. In contrast, from one year to the next, she might skip from grade 3 to grade 5, she might stay in grade 3, she could move down to grade 2, or she move to grade 4. This is stage structure.

3.1.1 An example of age structure

Consider a population where individuals live, at most, for five years and become mature at age 3. Skipping the usual process of deriving dynamic equations, instead we start with recursive equations to represent the dynamics of this population with a series of five equations:

$$N_{1,t+1} = F_3 N_{3,t} + F_4 N_{4,t} + F_5 N_{5,t} \tag{3.1}$$
$$N_{2,t+1} = S_1 N_{1,t} \tag{3.2}$$
$$N_{3,t+1} = S_2 N_{2,t} \tag{3.3}$$
$$N_{4,t+1} = S_3 N_{3,t} \tag{3.4}$$
$$N_{5,t+1} = S_4 N_{4,t} \tag{3.5}$$

Learn how to program this age-structured model in section 14.1.1 (pp. 213–21).

Where $N_{a,t}$ is the number of individuals of age a at time t, F_a is the number of offspring produced per individual aged a, and S_a is the proportion of individuals that survive from age a to age $a + 1$. The following advanced section (section 3.1.4) shows how these equations arise from dynamic equations.

The things that we might want to know about this population include the following:

1. Will the population size increase or decrease over time?
2. What is the relative abundance of each age (what is the distribution of ages in the population)?
3. Which life stages have the greatest reproductive value, and which model parameters contribute most to the population growth rate?

3.1.2 An example of stage structure

If the population is stage structured, we must account for individuals remaining in a stage for multiple time steps. We'll use the same notation as before, $N_{i,t}$, but now i refers to stage, not age. So $i = 0$ will mean offspring, $i = j$ will be juveniles that are not mature, and $i = a$ will be mature adults. We will dispense with the notation S_i for now because we are interested in not only how many individuals survive but also which stage the survivors are in. We use the notation G_i to denote the fraction of individuals in stage i that survive and move onto the next stage, and P_i to denote the fraction of individuals that survive and remain in the same stage.

For our hypothetical model, we'll assume that all surviving offspring transition into the juvenile stage, but individuals can survive and remain in the juvenile stage, and adults survive and remain in the adult stage. Our model then becomes

$$N_{o,t+1} = FN_{a,t} \tag{3.6}$$
$$N_{j,t+1} = G_oN_{o,t} + P_jN_{j,t} \tag{3.7}$$
$$N_{a,t+1} = G_jN_{j,t} + P_aN_{a,t} \tag{3.8}$$

In words, this says that the number of juveniles next year equals the number of offspring from last year that survived, plus the number of juveniles from last year that survived and remained in the juvenile stage. The number of adults next year equals the number of juveniles from last year that transitioned to become adults, plus the number of adults from last year that survived.

3.1.3   Transient and stable behaviors of structured models

Unlike their unstructured conterparts, structured models can exhibit "transient behavior" based on the initial age or stage distribution (figure 3.1). Depending on how one specifies the starting abundances in each age category, the initial population dynamics can be quite unusual at first. For example, consider the age-structured model above, where the initial number of reproductively mature individuals (ages 3 through 5) was high relative to the number of age 1 and age 2 individuals. Consequently, there will be a period of transient behavior where the population will initially grow very fast, but this

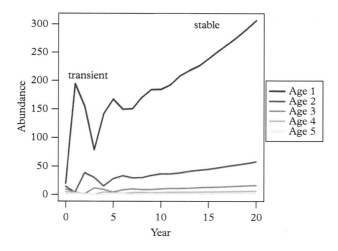

Figure 3.1 *Transient behavior of the age-structured model in equation 3.1 during the initial 10–15 years, which is replaced by stable behavior thereafter.*

cannot be sustained because it will take time for all of the newly created age 1 individuals individuals grow into mature individuals and produce offspring themselves. Eventually, however, these models will settle into a consistent pattern characterized by a *stable age distribution*, where the proportion of individuals at each stage is constant, and a *steady population growth rate*, where the total number of individuals next year is a constant fraction of the total number of individuals this year (figure 3.1).

3.1.4   Advanced: Deriving recursive equations

Underneath all of the models in this book are dynamic equations—equations that express the rate of change of state variables. We use those to derive recursive equations in discrete-time models that take the form $N_{t+1} = f(N_t) + N_t$, where $f(N)$ describes the rate of change in the state variable during the time step.

The previous section (section 3.1.3) showed recursive equations for an age-structured model without showing the underlying dynamic equations. This section will illustrate the linkage between the two more formally.

We start with $\frac{\Delta N_1}{\Delta t}$. New individuals come into this stage based on the fecundity of age 3, age 4, and age 5 individuals and on the number of individuals in those classes. Also, some age-1 individuals survive the year and become two-year-olds (this is a loss out of the N_1 state variable); we'll call these $S_1 N_1$. Equivalently, the number of individuals that die (also a loss) equals $(1 - S_1)N_1$. Because there are only two fates for individuals in the age 1 class die or move onto age 2, the sum of $S_1 N_1$ and $(1 - S_1)N_1$ must equal N_1. The dynamic equation for $N_{1,t+1}$ therefore equals

$$\frac{\Delta N_1}{\Delta t} = F_3 N_3 + F_4 N_4 + F_5 N_5 - N_1 \tag{3.9}$$

To make our recursive equation, we write

$$N_{1,t+1} = N_{1,t} + \frac{\Delta N_1}{\Delta t} \tag{3.10}$$

which equals

$$\begin{aligned} N_{1,t+1} &= N_{1,t} + F_3 N_{3,t} + F_4 N_{4,t} + F_5 N_{5,t} - N_{1,t} \\ &= F_3 N_{3,t} + F_4 N_{4,t} + F_5 N_{5,t} \end{aligned} \tag{3.11}$$

The rest of the ages are even simper. Take, for instance, the number of two-year-olds. Individuals enter this state from age 1, at a rate equal to $S_1 N_1$, and leave by either surviving and becoming three-year olds at the rate $S_2 N_2$, or by dying at rate $(1 - S_2)N_2$. Our recursive equation is therefore

$$N_{2,t+1} = N_{2,t} + \frac{\Delta N_2}{\Delta t} = N_{2,t} + S_1 N_{1,t} - N_{2,t} = S_1 N_{1,t} \tag{3.12}$$

Table 3.1 *Life table of the hypothetical population described in section 3.1.1*

Age	S_a	l_a	F_a
1	S_1	1	0
2	S_2	S_1	0
3	S_3	$S_1 S_2$	F_3
4	S_4	$S_1 S_2 S_3$	F_4
5	0	$S_1 S_2 S_3 S_4$	F_5

The same calculation can be performed for all subsequent age classes. The important point is that, despite the model being represented a recursive equation, there is an implied dynamic equation underneath it whose structure is completely consistent with our generalized dynamic model.

3.1.5 Advanced: Life-table analysis, reproductive outputs, and Euler's equation

Age-structured models are sometimes expressed in a "life table," which depicts the probability that an individual, once born, survives to any future age (usually denoted l_a), along with the age-specific survivorship and the age-specific reproductive output. In our example in section 3.1.1, we might create a life table that looks like table 3.1.

Notice that $l_1 = 1$, because of the way we defined one-year-olds: they are the first age being modeled. Because all individuals enter the population as age = 1, the probability of surviving to this age equals 1. The l_a terms can be written more succinctly for other ages as

$$l_a = \prod_{a=1}^{a-1} s_a \tag{3.13}$$

What does $\prod_{a=1}^{a-1}$ mean?

You've probably seen the notation $\sum_x$, which means sum all of the elements of x. The $\prod_x$ notation is similar but means multiply all of the elements of x together.

The expected lifetime reproductive output of an organism upon birth, R_0, equals

$$R_0 = \sum_{a=a_{\min}}^{a_{\max}} l_a F_a \tag{3.14}$$

where a_{min} is the youngest age in the modeled population, and a_{max} is the oldest age an individual can reach.

We can solve for the population growth rate of this population, λ, by applying the so-called Euler-Lotka equation, which states that

$$1 = \sum_{a=a_{min}}^{a_{max}} F_a l_a \lambda^{-a} \tag{3.15}$$

There is no easy way to algebraically manipulate this expression to solve for λ. Instead, you can use trial and error, or a numerical optimization method to find the value of λ that satisfies this equality. Note, however, that $\lambda = 1$ represents the case when the sum of F_a times l_a equals 1. In other words, it describes the case when, on average, each individual born will give rise to one offspring. It makes intuitive sense that the population would be stable if, on average, each individual replaces itself with just one offspring.

We sometimes express λ as $= e^r$, where r is the intrinsic growth rate (note here that $r = 0$ means no growth, $r < 0$ means population decline, and $r > 0$ means population increase), so you might see this equation expressed as

$$1 = \sum_{a=a_{min}}^{a_{max}} F_a l_a e^{-ra} \tag{3.16}$$

3.2   Modeling using matrix notation

It turns out that, for these types of models, there are simple mathematical solutions that are derived when you think of the models in terms of matrices. For students, matrices can often be intimidating, but that is more because they are unfamiliar than because they are complicated. Matrices simply allow us to write equations down more compactly.

Learn the basics of matrix multiplication in section 11.3 (pp. 185–86).

Define the matrix $\mathbf{A}$ and the vector $\mathbf{N}_t$ as follows:

$$\mathbf{A} = \begin{vmatrix} 0 & 0 & F_3 & F_4 & F_5 \\ S_1 & 0 & 0 & 0 & 0 \\ 0 & S_2 & 0 & 0 & 0 \\ 0 & 0 & S_3 & 0 & 0 \\ 0 & 0 & 0 & S_4 & 0 \end{vmatrix} \tag{3.17}$$

$$\mathbf{N}_t = \begin{vmatrix} N_{1,t} \\ N_{2,t} \\ N_{3,t} \\ N_{4,t} \\ N_{5,t} \end{vmatrix} \qquad (3.18)$$

so that the set of five equations used initially can be represented as

$$\mathbf{N}_{t+1} = \mathbf{A}\mathbf{N}_t \qquad (3.19)$$

Note that this looks a lot like our single-species model that had density-independent growth (equation 2.8). This is not a coincidence! Take a minute and confirm that this compact form gives us the same model that we started with above.

We can do the same thing with our stage-structured model (equation 3.6), where the matrix **A** is defined as follows:

$$\mathbf{A} = \begin{vmatrix} 0 & 0 & F \\ G_o & P_j & 0 \\ 0 & G_j & P_a \end{vmatrix} \qquad (3.20)$$

Notice now that the some of the diagonal elements of **A** include the P_i terms, to represent the individuals that survive and remain in the same stage from one time step to the next.

Learn how to program the matrix representation of the model in section 14.1.1 (pp. 213–21).

3.2.1   Why do we bother with matrix representation?

For one, it's quite a bit easier to write down a single equation instead of five. Second, mathematical manipulations of **A** tell us an awful lot about this population.

The matrix **A** is most generally referred to as a transition (or projection) matrix, because it includes information on the transfer of individuals from each age or stage to all other ages or stages. It is also sometimes called a Lefkovitch matrix. When the model is age structured (such as this one), then the matrix **A** is called a Leslie matrix. The features of the Leslie matrix are that the first row always has age-specific fecundities, survivorships are always listed in the off-diagonal, and all other elements are 0. When the model is stage structured, there can be nonzero elements anywhere in the matrix, depending on the life history of the species being modeled (Caswell 2000a).

3.3   Characteristics of structured models

An aside on eigenvectors and eigenvalues

Matrices that have the same number of rows as columns have some interesting properties. So-called eigenvalues and eigenvectors are one of these strange but useful properties.

For any matrix $\mathbf{A}$ that has m rows and m columns, there exists a value, called an eigenvalue (λ) and a $m \times 1$ vector, called an eigenvector ($\mathbf{v}$), that satisfies the following relation:

$$\mathbf{A}\mathbf{v} = \lambda\mathbf{v} \qquad (3.21)$$

This means there is a value, λ, that, when multiplied by the vector $\mathbf{v}$, gives the same output as the matrix multiplication of $\mathbf{A}$ and $\mathbf{v}$.

If that isn't strange enough, there are more than one pair of these eigenvalues and eigenvectors. Instead, there are m pairs of these eigenvalues and eigenvectors that satisfy this relationship.

In our example model above, we have a 5×5 matrix, so we have five eigenvalues, each with its own eigenvector. In structured population models, we only care about the largest of the eigenvalues and its corresponding eigenvector. We call these the *dominant* eigenvalue and eigenvector, respectively

You might be wondering how these are calculated. For small matrices, like a 2×2 matrix, it is pretty easy to do it by hand. For anything larger than that, things get difficult pretty quickly—no one in their right mind tries to calculate these manually. Unfortunately, spreadsheets usually do not have a function to calculate eigenvalues, but R and other software programs do this easily.

So what are these useful properties of matrices? Two important ones are the eigenvectors and eigenvalues of the matrix $\mathbf{A}$. Specifically, the dominant eigenvector $\mathbf{v}$ gives the stable stage distribution, that is, the relative proportion of individuals in each stage after the model has run for a while to remove artifacts caused by starting conditions. First, define $|\mathbf{v}|$ as $|\mathbf{v}| = \sum_i v_i$. That is, $|\mathbf{v}|$ is the sum of the s elements of the dominant eigenvector. Then, define $|\mathbf{N}_t|$ as the sum of the $\mathbf{N}_t$.

Using this notation, as t becomes large, the proportion of the population that is in each age or stage equals

$$\frac{\mathbf{N}_t}{|\mathbf{N}_t|} = \frac{\mathbf{v}}{|\mathbf{v}|} \qquad (3.22)$$

In other words, the so-called stable age distribution—the proportion of the population in each age—is given by the elements in the dominant eigenvector $\mathbf{v}$, divided by the sum of these elements (Caswell 2000a).

For simplicity, we'll refer to $\mathbf{v}^\star$ as the eigenvalue $\mathbf{v}$ divided by the sum of $\mathbf{v}$. Thus, returning to our population model, after several time periods have elapsed, the model will behave as

$$\mathbf{N}_{t+1} = \mathbf{A}\mathbf{N}_t = |\mathbf{N}_t|\mathbf{A}\mathbf{v}^* \tag{3.23}$$

But, because we know that $\mathbf{A}\mathbf{v} = \lambda\mathbf{v}$, and because $\mathbf{v}^\star$ is just a standardized version of $\mathbf{v}$, we can replace the right-hand side:

$$\mathbf{N}_{t+1} = |\mathbf{N}_t|\mathbf{A}\mathbf{v}^* = \lambda|\mathbf{N}_t|\mathbf{v}^* \tag{3.24}$$

How is this useful? Well, recall our initial density-independent model with no age-stage structure: $N_{t+1} = \lambda N_t$. Compare this to our equation above (equation 3.24). This shows us that the dominant eigenvalue λ gives us the population growth rate, completely equivalent to the population growth rate that we used in the simple model without age structure (Caswell 2000a).

Learn how to calculate eigenvalues and eigenvalues in section 14.1.1 (pp. 213–21).

3.4   Elasticity analysis

Another feature that we might want to know from the population is the reproductive value of each stage. The reproductive value of a stage is the expected contribution of individuals in this stage to future population growth. Put another way, individuals of some stages might contribute a lot to population growth because they stand a good chance of becoming sexually mature, producing viable offspring, or perhaps reproducing multiple times or having high fecundity. Consider a newly fertilized egg: it has relatively low reproductive value because it does not stand a good chance of surviving to maturity. Similarly, in our hypothetical example above (equation 3.1), an age 5 individual also has low reproductive value, because it will not live beyond the current year (so it has little contribution to future growth). In comparison, an age 2 individual in this model has high reproductive value because they have a good chance of reaching maturity and then reproducing for several years.

The vector of reproductive values is equal to the so-called dominant left eigenvector, which is calculated as the dominant eigenvector of the transpose of $\mathbf{A}$. This calculation accounts for the future fecundity and for the expected number of future reproductive events, as well as for the probability of surviving to each future reproductive event.

Ultimately, we would like to know which parameters are most important in determining population growth rate. One way to do this would be to change each parameter, one at time, and see the effect on population growth rate (λ). An elegant mathematical trick called elasticity analysis will calculate the proportional change in λ associated with a proportional change in each element of **A**.

The specifics of how to calculate this are presented in detail below, but you don't need to know these details to interpret elasticities. The important thing to know is that elasticities are calculated directly and analytically from the properties of the transition matrix.

So how and why would we use this? Well, it is obviously useful to know which of the parameters have a very strong influence and which do not. We might be interested in this from a purely academic standpoint, or perhaps a more pragmatic one. Which conservation actions are likely to be most effective? The answer to that question is complex, but a good rule of thumb might be "those that change influential demographic rates" (Gerber and Heppell 2004). A classic example is given by Crouse et al. (1987), who asked which of two conservation actions would be most effective to help loggerhead turtle populations. On one hand, turtles are captured incidentally in shrimp fisheries. These are mostly subadult and adult turtles. On the other hand, sandy beaches where eggs are buried and hatch are increasingly being developed for tourism and other human uses. Which conservation action would have the greatest benefit? By using elasticity analysis, it was clear that changing subadult mortality (i.e., reducing incidental take in shrimp fisheries) would have a far greater benefit to turtle populations than enhancing egg survivorship via beach habitat protection.

But elasticities must be interpreted with caution. The elasticity of any given parameter depends on the values of all of the other parameters. A slightly different parameterization might yield very different results. Also, when used for conservation actions, they only tell you half the story. They tell you which parameters are important, but they don't tell you which parameters are easily changed through conservation action. Caswell (2000b) distinguishes between retrospective and prospective elasticity analysis. The latter is what we've done here, which is looking at which parameter is driving the population forward. The former asks, "What parameters were most responsible for past changes in the real population?" The second question requires that we know how much the parameter values actually changed.

> Learn how to calculate elasticities in R in section 14.1.1 (pp. 213–21).

3.4.1   Advanced: How to calculate elasticities

We wish to determine the sensitivity of the population growth rate, λ, to each of the elements in the transition matrix **A**. To do this, we use the fact that matrices have both left and right eigenvectors. To be consistent with the original derivation by Caswell (2000a), we'll refer to **v** as the left eigenvectors, and **w** as the right eigenvectors (this is the opposite of what we did above, where **v** were our right eigenvectors). Thus, for any square matrix **A**, there are vectors such that

$$\mathbf{Aw} = \lambda\mathbf{w}$$
$$\mathbf{v'A} = \lambda\mathbf{v'} \tag{3.25}$$

where $\mathbf{v}$ can be calculated as the right eigenvector of the transpose of $\mathbf{A}$, $\mathbf{A'}$ (transpose means the rows and columns of the matrix are swapped).

The value of interest, $\frac{\delta\lambda}{\delta a_{ij}}$ (where a_{ij} is the element of $\mathbf{A}$ at the ith and jth position) equals

$$\frac{\delta\lambda}{\delta a_{ij}} = \frac{v_i w_j}{\langle \mathbf{w}, \mathbf{v} \rangle} \tag{3.26}$$

where v_i and w_j represent the ith and jth element of $\mathbf{v}$ and $\mathbf{w}$, respectively, and $\langle \mathbf{w}, \mathbf{v} \rangle$ is the scalar product of w and v, defined as

$$\langle \mathbf{w}, \mathbf{v} \rangle = \sum_{i=1} w_i v_i \tag{3.27}$$

The equation above then tells us the derivative of λ with respect to each a_{ij}. The elasticity gives the proportional change in λ with a proportional change in a_{ij}. This is denoted e_{ij}, and is calculated as

$$e_{i,j} = \frac{a_{ij}}{\lambda} \frac{\delta\lambda}{\delta a_{ij}} \tag{3.28}$$

All of the elasticities will sum to 1, so that each elasticity indicates the contribution of that parameter to the population growth rate.

3.5 Advanced: Structured density-dependent models

There is no reason that you cannot introduce density dependence into a structured population model. You need to ask at least two questions: "What parameters are density dependent?" and "What state variables do they depend upon?" There is no clear-cut guidance here, and you'll have to rely on your understanding of the population. Two common choices are as follows: assume that reproduction (either fecundity or fertility) is density dependent, and depends on the overall number of mature adults, or assume that survivorship through early life-history stages is density dependent and depends on the number of individuals in those stages. But, of course, you could place density dependence into every rate parameter in your model. You would end up with a model that is hopelessly intractable mathematically, but perhaps it is necessary for your population of interest and modeling question.

A true story

When I was a graduate student, I was trying to develop a two-stage (juvenile, adult) model of largemouth bass. I put Ricker-like density dependence in each stage. The model looked like this:

$$J_{t+1} = \alpha_a A_t \exp(-\beta_a A_t)$$
$$A_{t+1} = \alpha_j J_t \exp(-\beta_j J_t) + s_a A_t$$

I then spent the better part of two days trying to use algebra in the misguided hope that I could find an expression for the equilibrium abundances of juveniles J and adults A. That proved to be a hard-earned lesson that "not all models have analytical solutions," even seemingly easy ones. Sure, there *was* an equilibrium (there actually were one or three equilibrium points), but no amount of algebra would provide an expression.

One common assumption is to make reproduction density dependent and survivorship density independent. This is based on the idea that resource scarcity (food, habitat) is particularly important during early stages and reproduction (e.g., number of suitable nesting sites).

If we revisit the model shown in section 3.1.1 but wish to add density dependence to it, we might use the following model:

$$N_{a,t} = \begin{cases} f(N_{3,t}, N_{4,t}, N_{5,t}) & \text{if } a = 1 \\ N_{a-1,t-1} S_{a-1} & \text{otherwise} \end{cases} \tag{3.29}$$

where $f(N_3, N_4, N_5)$ is a density-dependent function that makes per capita reproductive output decrease as more potential offspring are produced. Let's denote the total reproductive output of adults at time t as E_t: these could be the number of fertilized eggs, for instance. We could then apply the Beverton-Holt equation (section 2.4.2) to model the number of age 1 individuals produced:

$$f(N_3, N_3, N_5) = \frac{\alpha E_t}{1 + \beta E_t}$$
$$E_t = \sum_a N_{t,a} F_a \tag{3.30}$$

Here, α is the maximum survivorship of initial offspring (e.g., eggs) into the first age class of the model, and the parameter β describes the density-dependent effects on reproduction. In many models, α will simply equal 1, as often the fecundity or fertility terms, F_a, account for this density-independent survivorship (in other words, $F_a = 10$

doesn't necessarily mean that ten eggs are fertilized but rather that ten offspring are produced that survive to the first modeled life stage).

Finding the equilibrium abundance is a bit tricky here, but, from section 3.1.5, we know that when the population is at equilibrium, the number of individuals at each age equals the total number of births (or, in this case, age 1 individuals) at equilibrium times l_a (the probability of surviving to age a). If we define B^* as the total number of age 1 individuals produced at equilibrium, then we can replace $N_{t,a}$ with $B^* l_a$. Here B^* is the total number of age 1 individuals produced at equilibrium. Therefore,

$$B^* = \frac{\alpha B^* \sum_a l_a F_a}{1 + \beta B^* \sum_a l_a F_a} \tag{3.31}$$

which we can rearrange and solve for B^*:

$$B^* = \frac{\alpha}{\beta} - \frac{1}{\beta \sum_a l_a F_a} \tag{3.32}$$

Once we know B^*, we can solve for the equilibrium abundance of each age class as $N_a^* = B^* l_a$.

Summary

- More detailed structured population models can be used to project population growth and population composition but are often done in a density-independent fashion for mathematical ease.
- When structured models are density independent, we frequently model them using a transition matrix, whose properties tell us the population growth rate and size structure.
- Structured models can also be made density dependent by making at least one rate depend upon at least one state variable.

Exercises

1. Use excel or R to code up the structured population model described in section 3.1.1 and run the population for 50 years, with some different starting population sizes. Apply the describe, explain, and interpret scheme to your model.

continued

2. Repeat the first exercise, but using matrix algebra. If using R, calculate the population growth rate and stable age distribution with the `eig` command.

3. Do a literature search on "population elasticity" and find an interesting paper that constructed and analyzed a structured population model. Read the paper, evaluate how well the paper documents its assumptions, and evaluate the author's attempt to "describe, explain, interpret."

4

Competition and Predation Models

4.1 Introduction

4.1.1 Consider the following two ecological scenarios

Rufous and broad-tailed hummingbirds compete with each other for access to nectar. The migrant and highly territorial rufous hummingbirds displace the resident broad-tailed hummingbirds. Broad-tailed hummingbirds and other nectar-feeding birds act as "nectar robbers," and aggressively challenge rufous hummingbirds for access to nectar (Kodric-Brown and Brown 1978).

Spiders eat aphids. The more spiders there are, the more aphids are killed and the more spiders are produced. The more aphids there are, the more aphids are produced. Aphid eat plants, and plant productivity is greatest when plants are abundant (Price et al. 1980).

Rufous and broad-tailed hummingbirds are natural enemies. Their coexistence seems implausible. Spiders are enemies of aphids, while aphids are enemies of plants, but spiders rely on aphids to survive, while aphids rely on plants to survive. When can we achieve some sort of balance between these feedbacks, and when do the feedbacks spiral out of control?

For both of these scenarios, it's hard to see the conditions that would allow each species to coexist. In other words, what needs to happen for there to be "stable equilibria" that contain all possible members of the system? This is where simple models can be useful: we can represent these interactions with mathematical equations and then solve for conditions that allow species to coexist.

Here we'll learn three techniques that allow you to take a model system and determine whether the system has a stable equilibrium with all members present.

4.1.2 What does "stability" mean?

When we talk about stability, we are talking about about the following:

- properties of equilibrium states
- what happens when there is some perturbation from the equilibrium state

Introduction to Quantitative Ecology: Mathematical and Statistical Modelling for Beginners. Timothy E. Essington, Oxford University Press. © Timothy E. Essington 2021.
DOI: 10.1093/oso/9780192843470.003.0004

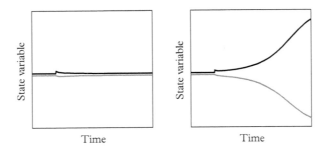

Figure 4.1 *Examples of a stable equilibrium (left) an unstable equilibrium (right). Stability is judged by the response of the system when perturbed from equilibrium.*

Put another way, we determine whether a model is stable by asking whether the *equilibria* are stable. An equilibrium is stable if a perturbation is followed by a return back to the equilibrium state. For instance, in the above hummingbird example, we can track the number of the two competing species (figure 4.1). On the left, the addition of a few extra individuals of one species in year 20 is followed by a return to the equilibrium state. *This is stable behavior.* On the right is the same model with different parameter values. Here, the addition of a few extra individuals of one species caused the model to move away from its equilibrium, where the species denoted by the black line outcompeted other. *This is unstable behavior.* It is unstable because when the system is perturbed, it reorganizes and drifts to a new equilibrium state.

4.2   Solving for equilibria

The easiest test is to find the equilibrium values of the state variables and determine the conditions under which all state variables have positive equilibrium values. For most ecological models, negative numbers are nonsense, so if any model predicts negative (or 0) abundance at equilibrium, we interpret this as evidence that the state variable cannot coexist with the other state variables.

For demonstration purposes, we'll consider an example of two species that compete with each other, such as the rufous and broad-tailed hummingbirds described above. We'll create a model that is a simple extension of the discrete-time logistic single-species population model. Denote the two species as X and Y:

$$X_{t+1} = X_t + r_x X_t \left(1 - \frac{X_t + \alpha Y_t}{K_x} \right)$$

$$Y_{t+1} = Y_t + r_y Y_t \left(1 - \frac{Y_t + \beta X_t}{K_y} \right) \tag{4.1}$$

where r_x and r_y are the maximum population growth rates of X and Y, respectively, K_x and K_y are the carrying capacities of X and Y, respectively, when competitors are absent.

The competition term α denotes the per capita effect of Y on X, relative to the effect of an X on X. In other words, it's a measure of the intensity of interspecific to intra-specific competitive effects. The parameter β is interpreted the same way. This model is often called the Lotka-Volterra competition model.

We solve for equilibrium by defining $X^\star$ and $Y^\star$ as the values of the state variables that make $X_{t+1} = X_t$ and $Y_{t+1} = Y_t$, and algebraically solving for the values of $X^\star$ and $Y^\star$ that satisfy these equalities. For the X part of the model, we get

$$X^* = X^* + rX^*\left(1 - \frac{X^* + \alpha Y^*}{K_x}\right) \text{ divide by } X^\star$$

$$1 = 1 + r\left(1 - \frac{X^* + \alpha Y^*}{K_x}\right) \text{ rearrange}$$

$$0 = \left(1 - \frac{X^* + \alpha Y^*}{K_x}\right) \tag{4.2}$$

We do the same for the Y part of the model:

$$0 = \left(1 - \frac{Y^* + \beta X^*}{K_y}\right) \tag{4.3}$$

We have two expressions and two unknowns, so we follow the usual algebra steps to solve for X^* and Y^*, yielding the two equilibrium expressions:

$$X^* = \frac{K_x - \alpha K_y}{1 - \alpha\beta}$$

$$Y^* = \frac{K_y - \beta K_x}{1 - \alpha\beta} \tag{4.4}$$

A savvy reader might note that there are three other equilibrium points: both X and Y can equal 0, X can equal 0 when $Y = K_y$, and X can equal K_x when $Y = 0$. However, here we are interested in the above case where the two species potentially coexisit.

Thus, we can define the sets of parameters K_x, K_y, α, and β for which $X^\star$ and $Y^\star$ are always greater than 0. This is the first step in identifying when species can coexist.

4.2.1 Continuous-time model

We use the same basic steps to solve for equilibrium if our model is in continuous time. The continuous-time version of the above model is

$$\frac{dX(t)}{dt} = r_x X(t)\left(1 - \frac{X(t) + \alpha Y(t)}{K_x}\right)$$

$$\frac{dY(t)}{dt} = r_y Y(t)\left(1 - \frac{Y(t) + \beta X(t)}{K_y}\right) \tag{4.5}$$

To solve for equilibrium, we set each differential equation equal to 0 (because this implies that neither state variable is changing) and use algebra to solve for X^* and Y^*. If you start this calculation, you'll notice that you are essentially doing the same algebra steps as above and will get the same equilibrium. This is because the steps applied in discrete-time models and continuoustime models are both doing the same thing: they are finding the conditions when the state variables are not changing.

4.3   Isocline analysis

The "equilibrium solution" approach works fine if the equilibrium is stable. As we saw earlier (section 2.5), some models have unstable equilibria. If you push them away just a little bit from an equilibrium point, they will drift away from that point, not back towards it. So, just knowing that the equilibrium values are positive doesn't tell you whether that equilibrium is stable.

As long as you have only two state variables, you can evaluate the stability of your model visually using isocline analysis. Consider a general discrete-time model of two competing species, X and Y:

$$X_{t+1} = f(X_t, Y_t)$$
$$Y_{t+1} = g(X_t, Y_t) \tag{4.6}$$

This model says that the dynamics of X and Y depend on levels of the two state variables, as described by the functions $f(X, Y)$ and $g(X, Y)$. To make this more tangible, let's assume that $f(X, Y)$ and $g(X, Y)$ are given by the Lotka-Volterra competition model in the previous section (equation 4.1), and use the parameters in table 4.1.

We begin by drawing a curve that shows the combination of X and Y that makes $f(X, Y) = X$. We call this an "isocline" because "iso" means "same," and when $f(X, Y) = X$, there is no change in state variable X from one time step to the next:

Table 4.1 *Parameter values used in the Lotka-Volterra competition model*

Parameter	Value
r_x	0.25
K_x	1
α	0.2
r_y	0.25
K_y	2.5
β	1.0

$$X_{t+1} = X_t + r_x X_t \left(1 - \frac{X_t + \alpha Y_t}{K_x} \right) \tag{4.7}$$

Setting $X_{t+1} = X_t$, dropping the t subscripts, subtracting X from each side, and dividing each side of the equation by r_x, we get

$$0 = X \left(1 - \frac{X + \alpha Y}{K_x} \right) \tag{4.8}$$

Notice that there are least two different ways this equality can be met, because the left-hand side consists of two terms multiplied together, X and $1 - \frac{X + \alpha Y}{K_x}$. If either of these terms equals 0, then the equality is met, so we must have two solutions. The first is when $X = 0$. This makes sense in a closed population (no immigration) because if there are no individuals of species X present, no new individuals of X can suddenly appear, so the population growth rate equals 0. Note that it doesn't matter how many Y are present; if $X = 0$, then $X_{t+1} = X_t = 0$. This particular solution is sometimes called a "zero isocline" or "null isocline" and is represented graphically by a line on the y-axis positioned at $X = 0$ (figure 4.2). The second solution happens when the second term equals 0, which will be a line showing the combination of values of X and Y that make the second term equal to 0 (figure 4.2).

Remember, the X-isocline shows the combinations of X and Y for which $X_{t+1} = X_t$. Ignore the zero isocline for the moment and look at the other isocline. If you are above this isocline, then there are too many Y or too many X, so $X_{t+1} < X_t$. Therefore, X will decrease until it hits the isocline. Similarly, if you are below this isocline, there are too few X and Y, so $X_{t+1} > X_t$. Therefore, X will increase until it hits the isocline. This interpretation is based on us knowing that this is a model of competition. In a competition model, adding more individuals of the same or different species will always reduce the population growth rate. If this were a predator prey model, say, where X preys upon Y,

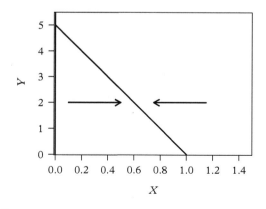

Figure 4.2 *The x isoclines show the x combinations of* X *and* Y *where* X *will grow (left) and shrink (right).*

then we would presume that more Y is beneficial to X. In that case, the area below the curve would imply an area where there was insufficient Y to support X, so X would decline, and vice versa.

The isocline therefore tells us quite a bit about the dynamics of the system. A key thing to remember when interpreting these is that *the X isocline only tells us about the movement of the model along the X axis* (in this case, horizontally).

We can now consider the isocline that shows where $Y_{t+1} = Y_t$, the Y isocline (figure 4.3).

Again, we focus on the downwards sloping isocline rather than the isocline that describes $Y = 0$. For points above the curve, there are too many X and Y so that $Y_{t+1} < Y_t$ and Y will therefore decrease. The opposite holds for points below the curve. Again, the Y isocline only tells us about the movement of the model in terms of the Y state variable (here, the vertical direction).

We can now combine these two graphs together to examine the overall change in the model's state variables (figure 4.4). Here, the dotted lines show the model movement in

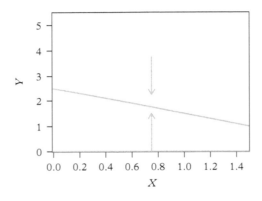

Figure 4.3 *The Y isoclines show the x combinations of X and Y where X will increase or decrease.*

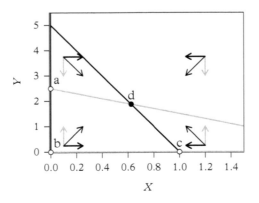

Figure 4.4 *We use the two-dimensional model movement for all different sections of the area bisected by the isoclines to predict stability of equilibria. Equilibria (circles, labeled a–d) are where the X and Y isoclines intersect each other. Solid circles depict stable equilibria.*

two dimensions. They are equivalent to force vector calculations that you might have done in physics.

4.3.1 Continuous-time model

The same basic steps apply to a continuous-time model. Presume we wish to use a continuous-time, recursive equation of the competition model, as shown in equation 4.5.

In a continuous-time model, the X part of the model is at equilibrium when $\frac{dX(t)}{dt} = 0$. To solve for the X isocline, we find the values of X and Y that makes the rate of change equal to 0:

$$\frac{dX}{dt} = 0 = r_x X \left(1 - \frac{X + \alpha Y}{K_x} \right)$$

$$0 = X \left(1 - \frac{X + \alpha Y}{K_x} \right) \tag{4.9}$$

We see again that when $X = 0$, the equality holds, because the left-hand side and the right-hand side both equal 0. Thus, $X = 0$ is one of the isoclines (the zero isocline). To find the other isocline, we divide both sides by X and rearrange to get

$$0 = r_x \left(1 - \frac{X + \alpha Y}{K_x} \right)$$

which, of course, is the same equality we discovered above (section 4.3.1) when calculating the isocline in the discrete-time model.

Learn more tips on calculating isoclines in section 14.2.1 (pp. 223–26).

4.3.2 Key points

- Isoclines are curves or lines that describe the combination of state variables that make one of the state variables constant.
- You can predict the model behavior along each axis, by seeing how that state variable will move based on the isocline position.
- The model direction is simply the combination of movements along two dimensions.
- Isoclines can be used for discrete-time or continuous-time models.

Now that we know what isoclines are, and how to read them, we can use them to identify equilibrium points and predict the stability of each.

4.3.2.1 Identify equilibrium

The isoclines for the state variable depict the combination of values of the two state variables that makes it unchanging. For the X state variable, any combination of X, Y

that lies on that line makes X constant from one time step to the next. For the Y state variable, any combination of X, Y that lies on that line makes Y constant from one time step to the next. The points where the isoclines intersect are where both $f(X, Y)$ and $g(X, Y)$ are not changing (i.e., this is a model equilibrium point). For the example we just used, there are four intersection points (figure 4.4), so we have four equilibrium points. Note the lower left-hand corner point is a trivial equilibrium where neither X or Y are present. Let's presume we are not interested in that one presently.

4.3.2.2   *Predicting stability*

Because the arrows show us the direction of model movement in all of the four regions that the two isoclines create (figure 4.4), we can look at each equilibrium point and see whether a small perturbation will be followed by a return to that equilibrium or instead by a move away from equilibrium. For example, take point d. If you perturb this equilibrium by adding a few extra Y (move up along the y-axis) and a few extra X (move right along the x-axis), the model will respond by countering that movement and return to point d. The same is true for all possible perturbations around that point. For this reason, we call point d *stable*, as the model wants to return to that state.

Consider instead point a. Do all of the arrows point back towards "a"? Nope, because the model will want to respond to a small perturbation, say, an addition of a few X, by increasing the abundance of species X and decreasing the abundance of species Y. Point a is therefore an unstable equilibrium. *If there is any section of the plot in the neighborhood of the equilibrium that points away from that equilibrium, then that equilibrium is unstable.* The same exercise will reveal that points b and c are unstable.

We conclude that the above model has a stable equilibrium where both species coexist and that equilibria that consist of just one of the two competing species are unstable. Is that always going to be true for this model? Let's try a different parameter set (table 4.2).

We calculate and plot the isoclines and find something different (figure 4.5). Here the intersection of the two sloping isoclines is unstable, because the arrows do not always

Table 4.2 *Alternative parameter values used in the Lotka-Volterra competition model*

Parameter	Value
r_x	0.25
K_x	1.5
α	0.75
r_y	0.25
K_y	3.5
β	4.0

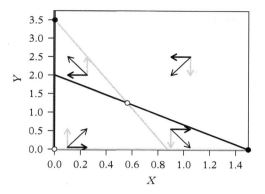

Figure 4.5 *Isoclines from an alternative parameterization of the Lotka-Volterra competition model.*

point back to the equilibrium point. Here, this model has two stable equilibria: one where only species X is present at its carrying capacity, and another where only species Y is present at its carrying capacity. Depending on the initial conditions, the model will eventually end up at one of these two stable equilibrium.

4.4 Analytic stability analysis

4.4.1 Motivation

The isocline method works when you have only two state variables. If you have more than two, is is difficult to visualize model stability in the same manner. Also, some models will have isoclines that don't obviously tell you about the model behavior. Indeed, many predator-prey models fall into this category. This happens because of the asymmetric effects of species on each other—predators are bad for prey, making prey less abundant, but prey are good for predators.

4.4.2 A brief aside on predator-prey models

One of the oldest predator prey-models is the Lotka-Volterra model, which represents the dynamics of a prey and predator pair, with the prey and the predator denoted N and P, respectively. These could be anything: hare and lynx, wildebeest and lion, or aphid and spider. The basic Lotka-Volterra model is

$$\frac{dN}{dt} = aN - bNP$$
$$\frac{dP}{dt} = cbNP - dP \tag{4.10}$$

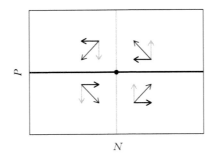

Figure 4.6 *Prey (black) and predator (gray) isoclines, with model trajectory arrows superimposed.*

where a is the intrinsic rate of population growth of prey, b is the attack rate of predators on prey, c converts prey eaten into new predators, and d is the death rate of predators. If you were to calculate isoclines, you'd see that state variables "disappear" from the isoclines (section 14.2.1), so that the isoclines are either horizontal or vertical (figure 4.6). When we draw the vector diagrams to predict model behavior, it is not immediately clear whether a small perturbation away from the equilibrium will result in the model returning to or moving away from equilibrium.

Learn how to simulate the dynamics of continuous-time models in section 14.5.

Another classic model is the Nicholson-Bailey host parasitoid model. Biologically, a host parasitoid system is different from predator-prey systems, in that parasitoids do not consume hosts. Instead, they lay eggs inside of their hosts, which then hatch and devour the host from the inside. The Nicholson-Bailey model works in discrete time, where the recursive equations are

$$N_{t+1} = \lambda N_t \exp(-aP_t)$$
$$P_{t+1} = cN_t(1 - \exp(-aP_t))$$

where λ is the host population growth rate when parasitoids are absent, a is the rate of search and capture, and c is the number of new parasitoids produced per infected host. The isoclines here are similarly ambiguous (figure 4.7) but, on first look, appear similar to those in the Lotka-Volterra model.

You might be surprised to know that these two models have rather different dynamic behaviors. The continuous-time Lotka-Volterra model is "neutrally" stable, meaning that, once perturbed, the model will oscillate around the equilibrium point, neither returning towards nor moving away from it (figure 4.8). Here there was a small perturbation so the cycles were very small around the equilibrium, but a larger perturbation would

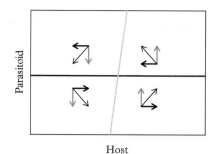

Figure 4.7 *Host (black) and parasitoid (gray) isoclines from the Nicholson-Bailey model.*

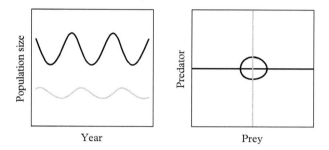

Figure 4.8 *Dynamic behavior of the Lotka-Volterra predator-prey model (black denotes prey, and gray denotes predator). The trajectory is shown in two ways: on the left, as population size versus time and, on the right, as pairs of N, P so that the dynamics can be shown on the same graph as the isocline. A small perturbation results in sustained oscillations that neither grow nor shrink in amplitude.*

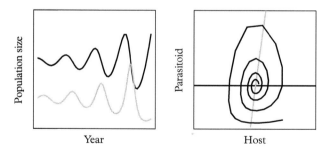

Figure 4.9 *Dynamic behavior of the Nicholson-Bailey host parasitoid model (black denotes host, and gray denotes parasitoid). A small perturbation leads to a spiral away from equilibrium.*

have created larger cycles. In contrast, the Nicholson-Bailey model is wildly unstable (figure 4.9), spiraling away from steady state no matter what parameters you use!

It would be great if we could predict these types of behaviors without simulating the model. Fortunately, there are some mathematical tricks that we can perform to identify whether an equilibrium point is stable and whether it will oscillate.

Describe, explain and interpret the Lotka-Volterra model

Describe: This model has feedbacks that lead to cyclic behavior, wherein the number of prey increases, then the number of predators increases, then the number of prey decreases, and finally the number of predators decreases. This cycle repeats itself, and the amplitude of the cycle depends on how far the initial abundances are from the equilibrium abundance.

Explain: The cyclic dynamics are a consequence of the asymetric interactions of species. Predators are bad for prey, so they make the number of prey go down, and prey are good for predators, so they make the number of predators go up. But there is always a lag in the population response, which causes cycles. In the Lotka-Volterra model, the stabilizing forces are exactly equal to the destablizing forces, creating a "neutrally stable" model.

Interpret: If real-world populations were only governed by these processes, then it would be hard to imagine how any population could maintain any semblance of equilibrium. That tells us that there are probably other feedbacks in real systems, like density dependence or other processes that are out of the model bounds (e.g., perhaps interactions with other species, or other environmental processes) that help stabilize these interactions.

4.4.3   Background and framework

Let's start with the rufous hummingbird–part of the general two-species competition equation (equation 4.1), and pretend like they don't interact with broad-tailed hummingbirds at all. Thus, $X_{t+1} = f(X_t)$.

By definition, the equilibrium value, X^*, occurs when $f(X^*) = X^*$. Suppose we consider a slight nudge away from X^*. That is, at some time point t, we'll push the system slightly away from equilibrium by an amount we'll denote x. Mathematically, we can relate X, x, and X^* as follows:

$$X_t = X^* + x_t$$
$$x_t = X_t - X^* \tag{4.11}$$

For the sake of stability analysis, we want to know whether the model will return to equilibrium or move away from equilibrium. *In other words, does the size of x_t get bigger or smaller over time?* If x_t grows through time, the perturbation is growing and therefore the model is unstable. If it shrinks through time, the perturbation is shrinking and the model is stable. We therefore need an expression for $\frac{\Delta x}{\Delta t}$. If this is less than 0, then the distance away from equilibrium will get smaller over time. If it is greater than 0, then the distance between X_t and the equilibrium will get larger over time.

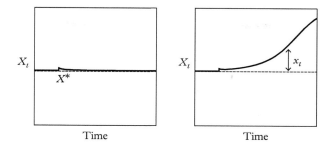

Figure 4.10 *Relationship between* X* *(dashed line),* Xₜ, *and* xₜ *in a stable (left) and an unstable (right) model.*

Using the same figure we used earlier in this chapter (figure 4.1), we can depict X^*, X_t, and x_t as in figure 4.10.

4.4.3.1 Key points

- Stability is defined as the rate of change of perturbations.
- Perturbations are defined as the difference between the state variable and the equilibrium.
- If the perturbations grow through time, then the equilibrium is unstable.

4.4.4 Calculating stability

We need a way to judge whether a perturbation will grow or shrink. Eigenvalues tell us this answer.

Return to the Lotka-Volterra competition model (equation 4.6) and presume that X and Y denote rufous and broad-tailed hummingbirds, respectively. Suppose we've introduced some perturbation, so both rufous and broad-tailed hummingbirds have been moved away from the equilibrium by an amount x and y, respectively. Will x and y shrink or will they grow?

4.4.4.1 Calculate the Jacobian matrix

The first step is to calculate the derivatives of $f(X, Y)$ with respect to both X and Y. Because X and Y are state variables that are themselves changing, we call these "partial derivatives," to denote that we hold the other state variable constant when performing the calculation. That is, when calculating the derivative of the function $f(X, Y)$ with respect to X, we treat Y as a constant, and denote this as $\frac{\delta f(X,Y)}{\delta X}$. All calculations are done at the equilibrium values, which are denoted X^* and Y^*, respectively. This gives us two derivatives: $\frac{\delta f(X,Y)}{\delta X}$ and $\frac{\delta f(X,Y)}{\delta Y}$. Repeat for the function $g(X, Y)$. The Jacobian matrix arranges these derivatives as follows:

$$\mathbf{A} = \begin{bmatrix} \frac{\delta f(X,Y)}{\delta X} & \frac{\delta f(X,Y)}{\delta Y} \\ \frac{\delta g(X,Y)}{\delta X} & \frac{\delta g(X,Y)}{\delta Y} \end{bmatrix} \tag{4.12}$$

This looks intimidating, but its just four different derivatives arranged into a single matrix.

4.4.4.2 *How do I know which derivative goes where?*

Good question! First, you need to decide the "order" of your state variables. It doesn't matter; you could order them as X then Y, or Y then X. The important thing is that you know the order that you're working with.

So think of the element in the first row and first column of the Jacobian matrix. It will be the derivative of the *first* state variable function, with respect to the *first* state variable. The element in the first row, second column, will be the derivative of the *first* state variable function with respect to the *second* state variable, and so on. In other words, the i, jth element of the Jacobian is the derivative of the ith state variable function with respect to the jth state variable. In the example above, I ordered X first and Y second. Had I reversed the order, making Y first and X second, my Jacobian would be

$$\mathbf{A} = \begin{bmatrix} \frac{\delta g(X,Y)}{\delta Y} & \frac{\delta g(X,Y)}{\delta X} \\ \frac{\delta f(X,Y)}{\delta Y} & \frac{\delta f(X,Y)}{\delta X} \end{bmatrix} \tag{4.13}$$

The following is *not* valid:

$$\mathbf{A} \neq \begin{bmatrix} \frac{\delta g(X,Y)}{\delta Y} & \frac{\delta f(X,Y)}{\delta X} \\ \frac{\delta g(X,Y)}{\delta X} & \frac{\delta f(X,Y)}{\delta X} \end{bmatrix} \tag{4.14}$$

Why is this wrong? The element in first row, first column, is the derivative of the Y state variable with respect to Y, so Y must be first, and X must be second. But this ordering is not maintained throughout the matrix. In the cell in the first row, second column, I have the derivative of the X state variable with respect to Y (the second state variable with respect to the first), which is the opposite of the correct order implied above.

4.4.4.3 *Calculate the eigenvalues of the Jacobian matrix*

Because $\mathbf{A}$ is a square matrix (here, 2×2), there are two sets of eigenvalues and associated eigenvectors. We only care about the eigenvalues, which are denoted λ_1 and λ_2, respectively. The largest of these tells you the rate of change of the perturbation. For discrete-time models, we look at the eigenvalue that has the largest "magnitude," as explained below, while, for continuous-time models, we look at which eigenvalue has the largest "real" part.

See an example calculation of the Jacobian matrix in section 14.2.2 (pp. 226–29).

4.4.4.4 *Interpreting eigenvalues*

Eigenvalues can have real and complex parts. For instance, the eigenvalue might be $-0.2 \pm 0.348i$. The complex part of the eigenvalue tells you something about the way

Table 4.3 *Using real and complex parts of the dominant eigenvalue to predict discrete-time model behavior*

	Magnitude <1	Magnitude >1
Complex part = 0	Smoothly move towards equilibrium	Smoothly move away from equilibrium
Complex part ≠ 0	Oscillate towards equilibrium	Oscillate away from equilibrium

Table 4.4 *Using real and complex parts of the dominant eigenvalue to predict continuous-time model behavior*

	Real part <0	Real part >0
Complex part = 0	Smoothly move towards equilibrium	Smoothly move away from equilibrium
Complex part ≠ 0	Oscillate towards equilibrium	Oscillate away from equilibrium

that it approaches equilibrium. If the complex part is 0 (e.g., there is no complex part), then the model will smoothly move towards or away from equilibrium. If the complex part is any number other than 0, then the model will oscillate either towards or away from equilibrium.

To determine whether the model will return or move away from equilibrium, you apply different calculations for discrete- and continuous-time models. For discrete-time models, you calculate the "eigenvalue magnitude" for each eigenvalue as $\sqrt{r^2 + c^2}$, where r is the real part, and c is the complex part. If the largest eigenvalue magnitude is less than 1, then the perturbations will shrink through time, and if it is greater than 1, the perturbations will grow through time (table 4.3). For continuous-time models, it is simpler: you simply look at the real parts of the eigenvalues and find the eigenvalue with the largest real part. If the value is less than 0, then the perturbation will shrink through time, and if it is greater than 1, then the perturbation will grow. Put more simply, the real value is the perturbation growth rate (table 4.4).

Many students ask me why the criteria for stability are different for the two types of models. For one, in a discrete-time model, we are attempting to estimate the discrete-time growth rate of x_t, which can be thought of as $x_{t+1} = ax_t$, where a is the discrete-time perturbation growth rate. Expressed this way, you can see why values of a greater than 1 imply growth of the perturbations. In a continuous-time model, we are attempting to estimate the continuous-time growth rate of $x(t)$, which can be thought of as $\frac{dx(t)}{dt} = ax(t)$. Negative values of the continuous growth rate parameter a imply a decay in the perturbations, while positive values imply growth.

A second difference is in the way that stability is calculated. In the continuous-time model, we only use the largest real part of the eigenvalues. In the discrete-time model, we calculate the eigenvalue magnitude, which includes contributions from both the real and the complex parts of the eigenvalues, and then choose the eigenvalue with the largest magnitude. You can think of the latter as a penalty for discrete-time models. Because discrete-time models do not constantly update rates in response to state variable values in the way that continuous models do, they are more prone to instability, especially when the model is oscillating.

4.4.4.5   *What about continuous-time models?*

Suppose your model was continuous, expressed as differential equations:

$$\frac{dX}{dt} = f\big(X(t), Y(t)\big)$$

$$\frac{dY}{dt} = g\big(X(t), Y(t)\big) \tag{4.15}$$

You would calculate the Jacobian matrix in the same way, but we interpret the eigenvalues differently as described above.

4.4.4.6   *Key points*

- We use calculus and linear algebra to represent the growth rate of a perturbation.
- When the perturbation growth rate is less than 1 (for discrete-time recursive models) or less than 0 (for continuous-time models), then the model is stable.
- When the largest eigenvalue has a nonzero complex part, the model will undergo oscillations.

4.4.5   Advanced: Why does the eigenvalue predict stability?

Because I find that the mathematical steps are slightly simpler in the continuous time model, we'll derive the basis for using Jacobian matrices and eigenvalues to determine stability. In a single state variable example, we define $x(t)$ as the difference between the value of the state variable at time t and the equilibrium value. If $\frac{dx}{dt}$ is negative, we know the model is stable because perturbations are shrinking. We need a way to calculate $\frac{dx}{dt}$ based on our model.

To begin, we recognize that $\frac{dx}{dt}$ must be related to the function $f(X)$ in some way. By definition,

$$\frac{dx}{dt} = f\big(X - X^*\big) \tag{4.16}$$

Because X^* doesn't change over time, this derivative must only be related to X, so that

$$\frac{dx}{dt} = f\big(X - X^*\big) = f(X) \tag{4.17}$$

and, by definition, $X(t) = X^* + x(t)$, so

$$\frac{dx}{dt} = f\left(X^* + x\right) \tag{4.18}$$

To derive an expression for $f(X^* + x)$, we need to use some sort of approximation. To summarize, $x(t)$ is the size of the perturbation. We have some function that describes the rate of change of X, the state variable. We need to use that function to calculate $\frac{dx}{dt}$.

4.4.5.1 *Using the Taylor approximation*

The Taylor approximation is a way to extrapolate the value of a function away from some known point. Specifically, we can approximate a function at some place $Z + \Delta z$ for any $f(Z)$ as

$$f(Z + \Delta z) = f(Z) + \frac{df(Z)}{dz} \Delta z \tag{4.19}$$

Understanding the Taylor approximation

Presume you knew sunset time today is 5:00 pm, and you wanted to predict sunset time ten days into the future. You also know that, right now, sunset time is changing at a rate of two minutes later per day. You would probably say, "I'll take today's sunset time, and add to it 2×10, giving me 5:20."

You just used a "first-order" Taylor approximation. You approximated some nonlinear function (sunset time vs. calendar date) into the future by taking the current value and the current rate of change and then linearly extrapolating.

This works in two dimensions as well. Suppose you want to know the sunset time ten days from now, in a location that is three degrees further south. And you know that, at this location and time of year, sunset time changes at a rate of 0.5 minutes per degree latitude. You would probably say "$5:00 + 2 \times 10 + 3 \times 0.5 = 5:21:30$." You did this by simply adding the current rate of change in the second dimension (in this case, latitude) into the calculation.

We can use this technique to derive an expression for $\frac{dx}{dt}$. Instead of Δz, we have $x(t)$, and, instead of $f(Z)$, we have $f(X)$. Also, $x(t) = X(t) - X^*$, so that $X(t) = x(t) + X^*$. We can substitute $x(t) + X^*$ in place of X in our function, $f(X)$, to get the value of $\frac{dx}{dt}$. This gives us

$$\frac{dx}{dt} = f(X^* + x) = f(X^*) + \frac{df(X)}{dX} x(t) \tag{4.20}$$

where the derivative $\frac{dX}{dt}$ is calculated at X^*. By definition, $f(X^*)$ equals 0, so this term goes away.

To simplify things, lets substitute a parameter in place of $\frac{d(f(X))}{dX}$. We'll denote this slope with the parameter a. With these changes, the expression becomes

$$\frac{dx}{dt} = f(X^*) + ax = ax \tag{4.21}$$

This equation, which keeps track of the size of the perturbation over time, looks exactly like an equation that we might use for density-independent population growth. If a is positive, the perturbations (the $x(t)$'s) will grow over time, which means that the system is moving away from equilibrium. If a is negative, the perturbations will get smaller over time, meaning that the system is moving closer to equilibrium. Thus, the stability of this simple system is determined solely by the parameter a, which is the derivative of the function $f(X)$ evaluated at equilibrium. For a single state variable, this is fairly intuitive: to be a stable equilibrium, dX/dt has to be negative when X exceeds X^*, and positive when X is less than X^*. This can only happen if $f(X)$ is a declining function of X near X^*.

If we had a discrete-time model, with a recursive equation, we would obtain

$$X_{t+1} = f(X) \tag{4.22}$$

and, to see whether a perturbation x_t is shrinking or growing, we would do the same steps but here ask whether the perturbation growth rate is less than 1. The approximation of this will simply be the slope of the recursive equation at equilibrium. If this slope exceeds 1, then the model will be unstable because this means our x_t will grow with each time step, much in the way our recursive density-independent population model grows when population growth rate exceeds 1.

Also, remember that, in section 2.5, we saw that the slope of the recursive equation at equilibrium governed the stability of the model. We can see that more clearly through the lens of the Taylor approximation. That is, we express the dynamics of the perturbation $x(t)$ as

$$x_{t+1} = ax_t \tag{4.23}$$

where a is the same as the parameter s we introduced in section 2.6.2: the slope of the recursive equation at equilibrium. You can now see why we get oscillations when s is less than 0, because the sign of x will change with each time step, flipping from positive to negative.

4.4.5.2 *Multiple state variables*

Using the same example and notation as we used earlier (equation 4.6), we have two differential equations to denote the dynamics of two state variables:

$$\frac{dX}{dt} = f(X,Y)$$
$$\frac{dY}{dt} = g(X,Y)$$

We assume that there is combination of X^*, Y^* for which $f(X^*, Y^*) = 0$ and $g(X^*, Y^*) = 0$. Once again, we'll consider a small perturbation of the system, this time adding a few extra individuals for each state variable:

$$X(t) = X^* + x(t)$$
$$Y(t) = Y^* + y(t) \tag{4.24}$$

The dynamics of $x(t)$ and $y(t)$ tell us whether the system is going to move closer to or further from equilibrium over time. We can use a first-order Taylor approximation again to get an expression for the dynamics of these perturbations, but this time it will be slightly more complicated, because the rufous hummingbird perturbation dynamics will also depend on the broad-tailed hummingbird perturbation:

$$\frac{dx}{dt} = f(X^*, Y^*) + \frac{\delta f(X,Y)}{\delta X}x + \frac{\delta f(X,Y)}{\delta Y}y$$
$$\frac{dy}{dt} = g(X^*, Y^*) + \frac{\delta g(X,Y)}{\delta X}x + \frac{\delta g(X,Y)}{\delta Y}y \tag{4.25}$$

Let's define the derivative of the f function with respect to X and Y as a_{XX} and a_{YX}, respectively. Similarly, define the derivative of the g function with respect to X and Y as a_{XY} and a_{YY}, respectively. Recognizing that $f(X^*, Y^*)$ and $g(X^*, Y^*)$ both equal 0, then we can rewrite this as

$$\frac{dx}{dt} = a_{XX}x + a_{YX}y$$
$$\frac{dy}{dt} = a_{XY}x + a_{YY}y \tag{4.26}$$

This is now looking a lot like a linear algebra problem, as each equation is a linear function of the state variables. Define the matrix $\mathbf{z}(t)$ as a vector of the perturbations $x(t)$ and $y(t)$:

$$\mathbf{z}(t) = \begin{bmatrix} x(t) \\ z(t) \end{bmatrix} \tag{4.27}$$

Now define a matrix, $\mathbf{A}$, that contains all of the slope coefficients:

$$\mathbf{A} = \begin{bmatrix} a_{XX} & a_{YX} \\ a_{XY} & a_{YY} \end{bmatrix} \tag{4.28}$$

This matrix is called the Jacobian matrix. We can now write the series of two equations as one compact equation:

$$\frac{d\mathbf{z}}{dt} = \mathbf{A}\mathbf{z} \tag{4.29}$$

If we define λ as the rate of change in $\mathbf{z}$, then

$$\lambda \mathbf{z} = \mathbf{A}\mathbf{z} \tag{4.30}$$

To solve for λ, we move all parts to the same side of the equation:

$$\mathbf{A}\mathbf{z}(t) - \lambda \mathbf{z}(t) = 0$$
$$(\mathbf{A} - \lambda \mathbf{I})\mathbf{z}(t) = 0 \tag{4.31}$$

where $\mathbf{I}$ is a 2×2 identity matrix (a matrix with 1s on the diagonal, and 0s elsewhere).

The set of equations that satisfy this equality are equal to the determinant of $\mathbf{A} - \lambda \mathbf{I}$, which is an m-order polynomial in λ, where m is the number of state variables. So, for a two-species model, the solution to this equation is a quadratic equation. The eigenvalues of $\mathbf{A}$ give the values of λ that solve the equality shown in equation 4.31. If we have a 2×2 matrix, we will have two eigenvalues. For any $m \times m$ matrix, we will have m eigenvalues.

The scalar λ is therefore the "perturbation growth rate," much in the same way that it relates to population growth rates. In this model, we have two dimensions, so we'll have two pairs of eigenvalues and eigenvectors. If all of the eigenvalues are negative, then this means that, over time, perturbations in all directions will eventually decline over time. If just one eigenvalue is positive, then perturbations will eventually increase over time. Thus, the condition for stability is that the largest real value of λ must be less than 0.

Had we modeled this with discrete-time recursive equations, the stability criterion would be different: the eigenvalues would all have to be less than 1. Remember, though, that we use the magnitude of the eigenvalues in discrete-time models. May (1973) explains this as a stability penalty for being a discrete-time model: discrete-time models adjust to changes in state variables more slowly than continuous-time models, so they are therefore inherently less stable.

4.5 Larger models

We close this chapter by illustrating how one can model entire food webs. Essentially, we want to expand the models developed here, particularly the predator-prey models, to include all of the feeding linkages in an ecosystem. These types of models are incredibly helpful for asking how structure and function are related. For instance, we might be interested in the cycling of limiting nutrients and how that contributes to ecosystem productivity. We might be interested in the structure of food webs, and how complex feeding linkages (such as intraguild predation, where a species both competes with and is a predator upon another species) affect stability. We might be interested in what processes contribute most to contaminant concentrations in species of interest. All of these require an broader approach, one not easily done by modeling a few interacting species.

Compartments models attempt to tackle this structural complexity. These models have a lot of breadth, but we know that when we add complexity along one axis (breadth), we sacrifice some complexity along another axis (detail). Consequently, compartment models usually make a few important simplifying assumptions:

- A system can be divided into a finite number of compartments; these are usually aggregations of species that play similar roles in food webs.
- Compartments are internally homogenous and connect to other compartments via the flow of energy or material.

Additionally, we often make functions that describe feeding linkages fairly simplistic. For instance, we might presume that the amount of prey eaten by predators depends only on prey abundance or that it depends only on predator abundance. These assumptions are clearly not universally "true" by any means, but they represent another set of convenient approximations that probably hold for certain conditions.

But, apart from these differences, large ecosystem or food web models are not structurally different from the two-species predator-prey models that we considered. Thus, you can apply the same procedures to build a model (here, thinking about rates that connect flow of material from one compartment to another, and what they depend upon), calculate equilibria values for all compartments, and evaluate the stability of these models. Readers interested in learning more about compartment models would enjoy the book by DeAngelis (1992).

Example

Tuna are large predators in the high seas and are among the most important species in fisheries around the world. At the same time, these species often have elevated mercury levels, so consumers are advised to moderate the amount of tuna they eat. Tuna in the eastern Pacific Ocean have higher mercury levels than those in the central Pacific Ocean. Are these differences due to oceanographic conditions that promote the conversion of mercury into a form that bioaccumulates? Or do they reflect differences in the structure of the food web? Specifically, because mercury bioaccumulates in food webs (meaning that concentrations get higher with each step in the food web), one hypothesis is that there are longer food chains in the eastern Pacific Ocean, where tuna mercury levels are highest.

Ferriss and Essington (2014) built models of how carbon and mercury move through each food web and then asked the following questions: are differences in mercury levels due to the underlying food web structure, and how does the food web structure dictate the amount of time mercury resides in the food web? The

continued

authors found evidence that both likely contribute: the only way to match empirical observations in the model was to make the mercury concentrations at the base of the food web larger in the Eastern Tropical Pacific. However, they also found that a sizable fraction of the difference between the two regions could be explained by food web structure. Thus, the model revealed that both mechanisms likely play a role in generating regional differences in mercury concentration.

Summary

- Many multivariable models are simple extensions of single-variable models.
- Models with multiple interacting state variables can be stable or unstable.
- Isoclines are a graphical way of evaluating stability: they depict the combination of state variable values that make one of the state variable constant; wherever isoclines intersect, you have an equilibrium point.
- Isoclines cannot work when there are more than two state variables, and they don't always tell you whether a model will oscillate.
- Analytical stability analysis tracks the perturbation growth rate; if the growth rate leads to a decay in the perturbation through time, then the model is stable.
- *Advanced*: Analytical stability analysis comes from a Taylor approximation of the model near equilibrium.

Exercises

1. Consider a model of competing species, X and Y, whose isoclines are depicted in figure 4.11. Which of the equilibria are stable, and which are unstable?

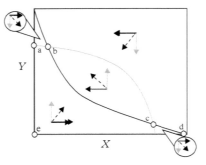

Figure 4.11 *The* X *isocline (black) and the* Y *isocline (gray) for a hypothetical competition model. Equilibria are denoted with circles.*

continued

2. Calculate the isoclines of the Lotka-Volterra predator-prey model (equation 4.10) and solve for the model equilibrium.

3. Another discrete-time competition model between two species, X and Y, might look like this:

$$X_{t+1} = \frac{\lambda_x X_t}{1 + a_x(X_t + \alpha Y_t)}$$

$$Y_{t+1} = \frac{\lambda_y Y_t}{1 + a_y(Y_t + \beta X_t)}$$

where λ is the maximum population growth rate, a_i controls the strength of density dependence, and α and β regulate the strength of interspecific competition. Use Excel or R to plot the isoclines and simulate the model trajectory. Choose model parameters based on your intuition of what seems sensible. See if you can switch the model from stable to unstable by changing the model parameters. What did you have to do to make the model unstable?

4. Calculate the Jacobian matrix of the Lotka-Volterra model and if you have access to R, calculate the eigenvalues using the `eigen()` function.

5. Code up and simulate the Nicholson-Bailey host parasitoid model, and confirm that no parameterization can make it stable. Apply the describe, explain, and interpret method to the Nicholson-Bailey host parasitoid model.

5

Stochastic Population Models

5.1 Introduction

The models that we've explored so far have been "deterministic." Their behavior is perfectly determined by the model equations and the starting values of the state variables. We know that real-world ecological systems don't behave that nicely. Instead, randomness (at least to our eye) governs ecological systems. We call this randomness "stochasticity." Thus, a stochastic model is one that explicitly includes randomness in the prediction of state variable dynamics. Because these models have a random component, each model run will be unique and will rarely look like a deterministic simulation (figure 5.1).

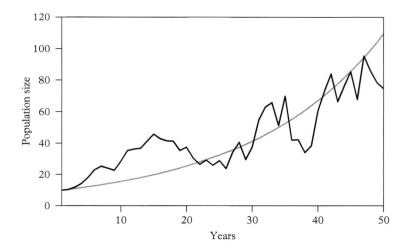

Figure 5.1 *Stochastic model runs (black) look different from deterministic model runs (gray) because the underlying rate variables are constantly changing.*

Introduction to Quantitative Ecology: Mathematical and Statistical Modelling for Beginners. Timothy E. Essington,
Oxford University Press. © Timothy E. Essington 2021.
DOI: 10.1093/oso/9780192843470.003.0005

5.1.1   What causes stochasticity?

We generally recognize two distinct sources that give rise to stochasticity. The first is probably the most obvious: properties of the ecological system (such as externally driven environmental conditions) add an element of randomness to state variables. We often call this "environmental stochasticity."

The second is more subtle, is most important for small population sizes, and takes a little bit of explaining. Recall that we often depict the number of surviving individuals from one point to another as the product of numbers at the beginning times the average survivorship. In reality, that survivorship probability applies to every individual in a population, as if, for each individual, there is a great existential coin toss that determines their fate. If it comes up heads, they survive; if it comes up tails, they do not. If you had 10,000 individuals, the expected number of survivors is 5,000. Even though in reality you might get 4,994 in one trial, and 5,015 the next trial, these deviations caused by the randomness of death rates are trivial in the grand scheme of things.

Consider instead a population of only ten individuals. While the expected number of survivors is 5, it would not be too strange if only two survived in a given year. Likewise, in other years, you might get eight, nine, or even ten survivors. At small population sizes, this randomness can really matter. The example in figure 5.2 shows one model run assuming a death rate (d) of 0.25 and that individuals give birth to one offspring annually with probability b, set to 0.3. Under a deterministic model, the population would show exponential growth at a rate of 5% per year, but any actual model run could be quite different, depending on the random chance events of births and deaths.

We call this second kind of stochasticity "demographic stochasticity" because it reflects randomness that results from demographic processes such as birth and death rates that must produce or remove entire individuals, not fractions of individuals. Naturally, this kind of stochasticity is most important when population size is low.

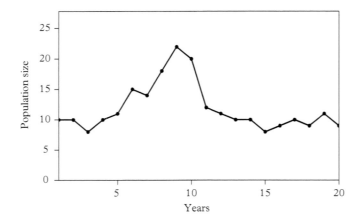

Figure 5.2 *When population sizes are low, randomness in the number of individuals that are born and die will generate dynamics that look different from deterministic models. Here, individuals give birth with a probability of 0.3 and have a 0.25 chance of dying, annually.*

5.2   Why consider stochasticity?

There are two main reasons why you might want to make your population models stochastic instead of deterministic.

5.2.1   Reason 1

Stochasticity leads to lower population growth than you would expect based on deterministic population growth. Consider our simple density-independent model: $N_{t+1} = \lambda N_t$.

To make the model stochastic, we assume that the parameter λ varies each year, using the notation λ_t:

$$N_{t+1} = \lambda_t N_t \tag{5.1}$$

Consider a 100 year trajectory of λ_t, first where λ_t is constant at 1.0 for all years t. The trajectory of N_t is unsurprising: population size remains unchanged throughout the 100 years (figure 5.3).

Now, add some variation in λ_t. Assume that λ_t varies randomly each year between 0.95 to 1.05. Each time we run this model, we'll get a unique trajectory of λ_t and therefore a unique trajectory of N_t, so we repeat this 100,000 times and calculate the average population size for each year. The result of this is shown in figure 5.3. Even though, on average, λ_t equals 1 for every one of the 100,000 simulations, the average population size in year 100 is about 2.5% less than it was when λ_t was held constant at 1. As we add more variation in λ_t, the effect is more pronounced, where we can reduce the average year 100 population size by nearly 15% by allowing λ_t to vary between 0.9 and 1.1 (figure 5.3).

Why does variation in λ_t reduce population size? One easy way to see this is to consider a simple case where we are projecting a model out two years from a starting population of 100: λ_t in year 1 is 0.9, and λ_t in year 2 is 1.1 (though the order does not matter). The average λ_t is 1.0. If we apply equation 5.1, we get

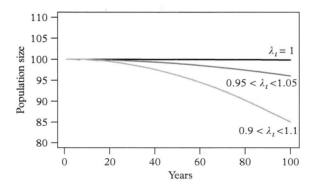

Figure 5.3 *Average yearly population sizes over 100,000 simulations for three different assumptions about variability in λ. When λ_t is allowed to vary between 0.9 and 1.1, the end population size is reduced by about 15%, compared to when λ_t is constant.*

$$N_{t+1} = N_t \lambda_t$$
$$N_{t+2} = N_{t+1} \lambda_{t+1}$$
$$N_{t+2} = N_t \lambda_t \lambda_{t+1}$$

and, by combining the first two expressions together,

$$N_{t+2} = 100 \times 0.9 \times 1.1 = 99$$

This says that even though the average λ_t was 1, the resulting population size was lower at the end than at the start. More information on why this happens is provided later.

5.2.2   Reason 2

Adding stochasticity allows you to look at interactions between population processes and external drivers to aid conservation and management decisions. For instance, one use of stochastic models is risk assessment, often called population viability analysis. Here you are less interested in the average condition of the state variable at some point in the future and more on the probability of reaching some minimum population size. This is commonly used in conservation biology to assess the prospects for recovering populations that have been pushed to low population sizes.

In fact, the vast majority of population models used in conservation and natural resource management are stochastic. For instance, fisheries scientists commonly use a stochastic age-structured model as the basis to estimate population abundance and to project forward the likely consequences of alternative fishing scenarios.

5.2.3   So, why aren't all models stochastic?

The reason that all models aren't stochastic is mainly because adding stochasticity adds another level of complexity to our model. Depending on your question, you may not be interested in accounting for the effects of stochasticity. Another cost of complexity is that only simple models have mathematical solutions for the mean and variance of predicted state variables through time. This means you often have to run models many times to numerically evaluate the range of future states. This takes time and may prevent you from seeing important underlying properties of your model. As a result, we typically include stochasticity when it is critical for answering our question.

5.2.4   Advanced: Why does stochasticity lower population abundance?

Consider an example using equation 5.1, where λ_t varies each year but in a somewhat contrived way such that there are exactly 50 good years, when λ equals 1.21, and 50 bad years, when λ equals 0.81. The arithmetic mean of λ is 1.01, suggesting slight population growth, but as you can expect from the previous section, the true population trend will be quite a bit different.

We first can generalize the expression to project population size any $t + 1$ years into the future, given a starting population size N_0:

$$N_{t+1} = \lambda_0 \lambda_1 \lambda_2 ... \lambda_t N_0 \tag{5.2}$$

Now, place the annual λ into the above equation, noting that 50 of the λ_t equal 1.21, and 50 equal 0.81. The equation simplifies to $N_{100} = (1.21)^{50}(0.81)^{50} N_t = 0.366 N_t$.

In other words, the population size 100 years from now will be roughly one-third of what it is now!

5.2.5 Why was the arithmetic mean incorrect?

It is because we used the "wrong" mean. The arithmetic mean is calculated by adding the λ_t over all of the years and dividing by the number of years. But we see above that the λ_t are not added together—they are multiplied together. In fact, we can see generally that

$$N_{t+1} = N_0 \prod_{i=0}^{t} \lambda_i \tag{5.3}$$

where the notation $\prod_{i=0}^{t} \lambda_i$ means to mulitply all of the λ_t together. We can work backwards to figure out the long-term average growth rate. Assume that there is some average growth rate that describes the overall trend, and call this $\bar{\lambda}_G$, so that the population size in year $t + 1$ equals

$$N_{t+1} = \bar{\lambda}_G^{t+1} N_0 \tag{5.4}$$

We want to solve for $\bar{\lambda}_G$, so we rearrange to obtain

$$\bar{\lambda}_G^{t+1} = \frac{N_{t+1}}{N_0} \tag{5.5}$$

We have an equation for N_{t+1} that we can substitute, so the right-hand side equals

$$\bar{\lambda}_G^{t+1} = \frac{N_0 \prod_{i=0}^{t} \lambda_i}{N_0} = \prod_{i=0}^{t} \lambda_i \tag{5.6}$$

We can therefore get $\bar{\lambda}_G$ from

$$\bar{\lambda}_G = \left(\prod_{i=0}^{t} \lambda_i \right)^{\frac{1}{t+1}} \tag{5.7}$$

The expression on the right-hand side is called the geometric mean. The geometric mean (in this case) when λ_t varies is always less than the arithmetic mean. This is known as Jensen's inequality. The difference between the geometric mean and the arithmetic mean increases as the variance of λ_t increases.

Advanced: Jensen's inequality

Suppose you have some function $f(x)$, and x is a random variable. Jensen's inequality states that the mean of $f(x)$, $\overline{f(x)}$, usually does not equal $f(\bar{x})$. That is, you usually don't get the average value of $f(x)$ by swapping in the average of x for x. This would only be accurate in the special case when $f(x)$ is a linear function of x. If $f(x)$ is a decelerating function of x, then $f(\bar{x})$ will be too big; if $f(x)$ is an accelerating function of x, then $f(\bar{x})$ will be too small. In our case, $f(x)$ is $\log(x)$, which is a decelerating function of x.

The principle is named after Johan Jensen, who provided the definitive treatment of it in 1906.

5.3   Density-independent predictions: An analytic result

Assume we are interested in projecting out three years with three known values of λ, λ_0, λ_1, λ_2, and a starting population size of N_0.

Those of you that read the previous section will know the underlying derivation, but, for everyone else, you can take at face value that population size in the final year is given by the geometric mean of the λ_t's (the geometric mean of t samples equals the product of the t samples, raised to the $1/t$ power). We call this geometric mean $\bar{\lambda}_G$. Thus, $N_3 = \bar{\lambda}_G^3 N_0$.

The geometric mean of λ to the third power equals the product of λ_0, λ_1, and λ_2: $\bar{\lambda}_G^3 = \lambda_0 \lambda_1 \lambda_2$.

Suppose you take the logarithm of each side:

$$3\log(\overline{\lambda_G}) = \log(\lambda_0) + \log(\lambda_1) + \log(\lambda_3) \tag{5.8}$$

and then calculate the arithmetic mean of the $\log(\lambda_t)$:

$$\log(\bar{\lambda}_G) = \frac{\log(\lambda_0) + \log(\lambda_1) + \log(\lambda_2))}{3} = \overline{\log(\lambda)} \tag{5.9}$$

At first, this does not look particularly helpful, but it actually reveals a useful notation shorthand. If the geometric mean of a series λ_t gives the correct population size in the final year, *and* the geometric mean is calculated from the arithmetic mean of the log λ_t's, then why not work in log space? That is, if you define r_t as $\log(\lambda_t)$, and $\bar{r}$ is the average of r_t, then the number of individuals in time step 3 will equal

$$N_3 = e^{3\bar{r}} N_0 \tag{5.10}$$

Students commonly ask, "Why do we bother working in log space, only to undo that by exponentiating?" I get it—it seems like an unnecessary step at first glance. You don't need to follow the detailed mathematical reason—just remember that (1) we like using arithmetic means, as it is a very intuitive way for us to think about numbers; (2) the population growth rate depends on the geometric mean, which is a nonintuitive way to think about numbers; and (3) taking the log of the annual population growth rates allows us to use the arithmetic mean. Think of it this way: if I told you the arithmetic mean of $\log(\lambda_t)$ equaled 0.01, you would quickly realize that, on average, this population has a population growth rate of $e^{0.01} = 1.01$, that is, population growth. If I told you that the arithmetic mean of λ_t equaled 1.01, you would have no way of knowing whether the population was growing or shrinking, because you would also need to know the amount of variability in the λ_t's.

Which logarithm should I use?

You probably learned that the logarithm of some number, x, was the power that you had to raise a base number, almost always 10, to get x. In other words, by default, the notation $\log(x)$ meant to find the base 10 logarithm. You probably also learned that you could use Euler's number as the base, which you called the natural log, which was denoted $\ln(x)$, or perhaps $\log_e$.

Mathematicians usually use $\log(x)$ to refer to the natural logarithm, or x. This is because Euler's number arises in many mathematical solutions (and we used it for some of our models, like the Ricker population model in section 2.4.2). Throughout this book, $\log()$ will always refer to $\log_e$.

5.3.1 Projecting forward with unknown future stochasticity

The above examples had known values of $\log(\lambda_t)$, which was useful to see how we might structure a stochastic population model. Now consider the case when we are projecting a model forward when we don't know the actual λ_t but can describe the probability distribution from which they are generated.

Presume that the r_t are drawn from some probability distribution, such as a normal distribution. If the normal distribution has mean μ and variance σ^2, then we can approximate the expected population size any t steps into the future as follows (Heyde and Cohen 1985; Tuljapurkar and Orzack 1980):

$$\mathbb{E}[N_t] = N_0 e^{\mu t} \tag{5.11}$$

Random variables

Note here that we are jumping ahead a little bit to consider random variables. In my experience, most people are familiar with a "normal distribution," even if they are unfamiliar with the math behind it. If you wish to jump ahead to see the math, read chapter 7; otherwise, just use your existing expertise.

This expression is biased when σ is large. Dennis et al. (1991) provide a more accurate expression:

$$\mathbb{E}[N_t] = N_0 e^{(\mu+\sigma^2/2)t} \tag{5.12}$$

These two expressions tell us the mean, or expected value, of N at any time t steps into the future, but what about the variance in N_t? Heyde and Cohen (1985) and Tuljapurkar and Orzack (1980) showed that the variance in $\log(N_t/N_0)$ equals $\sigma^2 t$. Translating this into variance in N_t is a bit messy, but we don't really need to bother at this point; the take-home point is that the variation in possible population sizes grows as we predict further out into the future. This is illustrated in a simple example of twenty simulations run over forty years (figure 5.4). The population sizes at $t = 10$ are fairly similar. But when we try to predict population size at $t = 40$, there is a huge range of possible outcomes.

Learn how to program your own stochastic model in section 14.4 (pp. 235–43).

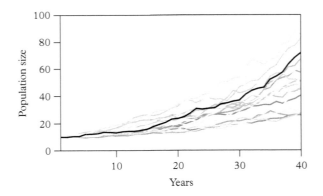

Figure 5.4 *Twenty different outcomes of the stochastic population model. Each line is a separate run. In stochastic models, we're interested in the mean and the variance of projected population size at any given time.*

What about continuous-time models?

One can also introduce stochasticity to continuous-time models, but we do not treat that topic here because the math is somewhat more complex, and also because it requires applying a fair amount of probability theory, which we do not begin until chapter 7 (Mangel 2006).

5.4   Eastern Pacific Southern Resident killer whales

We'll explore applications of stochastic population models using the Eastern Pacific Southern Resident killer whales (*Orcinus orca*). This species is listed under the US endangered species act as "Endangered," and it faces a number of threats. Things started to go badly for it in the 1960s and early 1970s, when juveniles were captured and placed in aquatic theme parks. Now it faces habitat loss, exposure to toxic contaminants, loss of prey, and a cacophony of noises from shipping vessels and other sources. Its populations are very diminished and are naturally small to begin with. While the plight of these whales is saddening, it provides a good test case for exploring the use of stochastic population models to estimate extinction risk.

We have very good estimates of population sizes for a number of years, because each individual can be identified. This is important, because the population estimates are a true census—scientists can keep track of every individual throughout their life. The scientists know who has lived, who has died, who has given birth, and so on. Surveys have been done annually since 1974 (figure 5.5).

We begin by asking the most rudimentary question regarding population risk: is there evidence that the population growth is positive? The simplest approach is to estimate

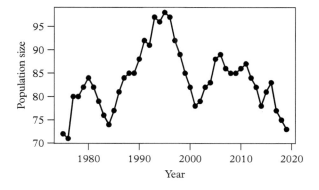

Figure 5.5 *Annual population estimates of killer whales. Data courtesy of the Center for Whale Research, https://www.whaleresearch.com/orca-population*

the average population growth rate, μ, by first calculating the r_t for each year as the ratio of $\log(N_{t+1}/N_t)$ and estimate the sample mean, $\bar{r}$. If we assume these are independent samples, we can use the usual methods to calculate the sample standard deviation, the standard error of the mean, and the associated confidence interval. For these data, the sample mean of the r_t is 0.0003 yr^{-1}; in other words, a small but positive growth rate. Yet, the sample standard deviation is pretty high, 0.038. The confidence interval around the mean is not encouraging: the 95% confidence interval is $(-0.011, 0.011)$, though note that, later (section 5.4.1), we see that the statistical assumption that each r_t is independent is likely violated, so don't take too much stock in these values. So we can't safely rule out that the long-term trend in the population is negative!

When we speak of extinction and risk, we typically aren't thinking about the literal event when the population disappears forever. Things start getting bad for populations well before then. At small population sizes, populations can experience a "extinction vortex," where the small population size makes other risks bigger (Gilpin and Soule 1986). For instance, once you're down to two individuals, you had better hope that they are of opposite sex. Also, if there are only two parents, the parents and all of the offspring will be closely related. This causes genetic inbreeding that reduces population fitness. Further, if there are only two parents, and they happen to put a nest in *that* tree, and *that* tree happens to get hit by lightening, that will be the end of that population. These threats are obvious at population size at 2, but they are also important at larger population sizes. We call this the "extinction vortex" because as the population size gets smaller, these other effects kick in and make the situation worse, causing further reductions in size. Rather than concern ourselves with the literal extinction event, we often instead focus on the risk (probability) that the population will reach some population threshold when the above issues start to occur. We often call this "quasi-extinction risk."

5.5　Estimating extinction risk

Even populations with positive population growth rates may go extinct. Think of it this way. Suppose it is 2019 and there are seventy-three killer whales. Further, suppose our extinction threshold—the point below which extinction risk raises rapidly—is forty killer whales. We want to know the probability that the population will drop below forty individuals next year. For this to happen, $\lambda_{t+1=2017}$ would need to be less than or equal to $40/73 = 0.55$. How likely is this? We calculate the risk assuming the $\log(\lambda_t)$ are draws from a normal distribution (with a mean of 0.0003 and a standard deviation of 0.038). Under these assumptions, the chances of the number of killer whales dropping below 40 is pretty small: 1.23×10^{-55}. Very, very small.

That calculation was for just one year. Now lets ask, "What is the probability of this happening either next year *or* the year after?" We want to know all of the ways that the product of $\lambda_{2019} \times \lambda_{2020}$ is less than or equal to 0.55. With two years, there are a lot more possibilities. The probability is still probably going to be small, but it will certainly be larger than 1.23×10^{-55}. By this same logic, think about 100 years, and all the ways the product of λ_t might, at some point, equal 0.55. Plus, we are interested whether the population *ever* crosses 40. We might get a string of λ_t draws that, when multiplied together, exceed 0.55 (let's say from year 1 to 100), but, just by chance, the product of the first 50 λ_t draws might be less than 0.55, meaning that it crossed the quasi-extinction threshold.

The probability of crossing the extinction threshold will depend on four things:

1. How far is the current population size from the quasi-extinction threshold? If we are far from it, the risk will be lower than if we are already right on the cusp.

2. Is the population growth rate moving the population away from the quasi-extinction threshold, and if so, how fast? A rapidly growing population is less likely to drop below the exception threshold.

3. How variable is the population growth rate? A population with highly variable population growth rates is more likely to have one or more bad years that pushes it below the extinction threshold.

4. How far out in the future are we projecting? As we saw above, the risk of extinction next year for killer whales is small, but the risk over 100 years is larger.

Dennis et al. (1991) worked out a mathematical expression for our population model (equation 5.11) that calculates the probability of extinction as a function of these factors. For simplicity, first think about the long-term extinction probability: as we project out to a very long time (infinite) horizon, what is the chance that we cross the population threshold? That probability is given by

$$P(N_{\text{extinct}} | \mu, \sigma^2, N_0 = N) = \begin{cases} 1, & \text{if } \mu \leq 0 \\ \frac{N}{N_{\text{extinct}}}^{-2\mu\sigma^2} & \text{if } \mu > 0 \end{cases} \tag{5.13}$$

The first expression simply says that if the population growth is negative, then you will always (eventually) cross the extinction threshold. The second expression says that even if population growth is positive, you might still cross the extinction threshold, owing to chance events. The second expression is also commonly written as: $e^{-2\mu x_d/\sigma^2}$, where x_d is the $\log(N/N_{\text{extinct}})$.

Now we can apply this method to analyzing the Southern Resident killer whales. Remember, we use the sample mean as our estimate of μ, which, in this case, exceeds 0, so we use the bottom expression. Using $N_0 = 73$, $N_{\text{extinct}} = 40$, $\bar{r} = 0.0003$, and $\sigma = 0.038$, the risk of eventual extinction is 0.77, or a 77% chance. Of course, this calculation depends on the choice of the quasi-extinction threshold. If the threshold is lower, say, 30, then the long-term risk is less: 0.66.

These calculations tell us the long-term extinction risk. What about shorter time frames? What is the risk of extinction from now to fifty years into the future? Ten years into the future? Did Dennis et al. (1991) figure out a mathematical expression for those? Yes, but the equation is pretty long and convoluted (the equation is so impressive, it appears on the cover of the book by Morris and Doak [2002]).

Alternatively, we can use Monte Carlo simulation to evaluate extinction risk by simulating the model for a set number of years, repeating that simulation several times, and evaluating in how many of the simulations did the population drop below the quasi-extinction threshold. Suppose we were interested in the probability of extinction over the next 100 years. For these model parameters, the Monte Carlo simulation indicated extinction would occur 9% of the time, or odds of roughly 1 in 11 (table 5.1).

Learn more about Monte Carlo simulation methods in section 14.3 (pp. 229–35), and the specific application to extinction risk in section 14.4 (pp. 235–43).

5.5.1   Advanced: Autocorrelation

The population risk can be more pronounced when there is autocorrelated variability in the rates. Autocorrelation means that good years are followed by good years, and bad years are followed by bad years. Think of weather in your location. When a weather system arrives, it affects the weather for multiple days. So, you may have several days of lovely weather followed by several days of bad weather. The same holds for populations and longer timescales. If one year is an especially good year, that might signal some broader environmental condition that persists for multiple years. As a result, the following year might have strong or poor population growth as well.

We can see some vague signs of autocorrelation in the killer whale data (figure 5.6). Notice the runs of above average and below average growth rates, occurring in 4- 10 year periods. For instance 1979 was a bad year, as was 1980, 1981 and 1982. That was followed by three years of positive population growth. There was another run of bad years starting in 1996 through 2000, which was again followed by four good years.

A common way to model this type of autocorrelation is with the following expressions:

$$r_t = \mu + \eta_t$$
$$\eta_t = \rho \eta_{t-1} + v_t$$
$$v_t \sim N(0, \sigma^2) \qquad\qquad (5.14)$$

Here ρ is the autocorrelation coefficient, and η_t is the difference between each year's r_t and the mean of the r_t. When the coefficient is 0, the r_t are independent. When ρ is 1, the process becomes a random walk, and intermediate values of ρ give intermediate degrees of autocorrelation.

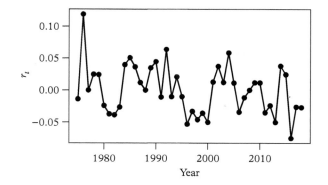

Figure 5.6 *Annual population growth rates suggest autocorrelation.*

One problem with this formulation is that the overall variance of r_t will depend on ρ. The following version will eliminate this dependence:

$$r_t = \mu + \eta_t$$
$$\eta_t = \rho\eta_{t-1} + \sqrt{1-\rho^2}v_t$$
$$v_t \sim N(0, \sigma^2) \tag{5.15}$$

For the killer whale data, ρ is around 0.23: how does this affect the estimated population risk? There is no equation to help us out here. Instead, we calculate extinction risk by running the model over and over again, keeping track of the number of model runs where the population dropped below the extinction threshold. That is, we run the model described by

$$N_{t+1} = e^{r_t}N_t \tag{5.16}$$

but generate r_t by using equation 5.14. I simulated 10,000 different model trajectories, each of 100 years, and counted the number of model trajectories where N dropped below the extinction threshold. If I use the $\rho = 0.22$ value, the extinction risk was 17%, almost double the rate when we assumed that the r_t were independent of each other.

Learn more about how to generate autocorrelation in these models in section 14.4.3 (pp. 239–42).

Why does autocorrelation do this? Suppose that r_t takes only "good" ($r_t = 0.05$) or "bad" values ($r_t = -0.05$), that r_t averages 0, and that the population is 20% above its quasi-extinction threshold. If good and bad years alternate perfectly, the population will simply oscillate up and down without any trend. If, instead, there are five bad years followed by five good years, the population will decline by over 20% before the good years arrive.

5.6   Uncertainty in model parameters

The above examples presumed that we knew μ and σ (and ρ) perfectly from the data. Of course, we only have the sample mean and the sample standard deviation, which may or may not be correct. Indeed, if the population growth rate is negative—a possibility that the confidence interval cannot rule out—extinction is guaranteed. An assessment of extinction risk ought to consider this uncertainty as well.

We can once again use Monte Carlo simulation, which will operate something like this:

1. We randomly sample a normal distribution to get a μ, and sample from another distribution to get a σ, both based on the data.
2. Based on this combination of μ and σ, we use equation 5.16 to project forward the population 100 years. For each projection, we note whether the population ever dropped below the extinction threshold.
3. We repeat the process over and over again and then calculate the proportion of times that the model population was less than the extinction threshold.

What people find most confusing is that randomness appears in the model twice. We have it in our usual way, in describing interannual variation in r_t, where each year's r_t is a random draw. But we also have it at a higher level, where the mean and the standard deviation of the r_t are also random draws for each model run.

Learn how to propagate uncertainty in stochastic models in section 14.4 (pp. 235–43).

Once we include the uncertainty in the model parameters as described above, the estimated extinction probability is really high—about 0.22, or a 1 out of 4 chance. Note that this does not include the effects of autocorrelation, which would make the probability even higher.

Table 5.1 summarizes the findings under different assumptions. Two big things are clear. One, the underlying assumptions have a large bearing on the estimated population risk. The analytic solution that assumes that we knew the parameters perfectly and that r_t are independent likely gives an optimistic estimate of extinction risk. The second main, take-home point is that, despite the fact that the killer whale population trajectory doesn't show a steep decline, the extinction probability for this population is substantial.

Table 5.1 *Comparison of killer whale extinction probability estimates over the next 100 years*

Model	Extinction probability
Parameters known, no autocorrelation	0.09
Parameters known, autocorrelation	0.17
Parameters unknown, no autocorrelation	0.22

5.7 Density-dependent stochastic models

Just as we extended a density-independent model to include stochasticity, we can do the same with density-dependent models. We need to take some care, as it is usually useful to keep birth and death processes separated and avoid models that have clever parameterizations for ease of interpretation (like the logistic). For instance, suppose you wanted to capture density dependence by making stochastic the intrinsic rate of population growth (r) in the logistic model. If the population size equals K, then, no matter what value r takes, the population will be unchanged, even with large stochastic changes in r. However, if you instead model the stochastic effects of births, deaths, or both, you can avoid this strange outcome.

For instance, one could use a density-dependent model of births and a density-independent model of deaths (e.g., equation 2.19) and add stochasticity to one or both of these:

$$N_{t+1} = N_t + N_t \left(\frac{\alpha_t}{1 + \beta N} - d_t \right) \tag{5.17}$$

where the maximum birth rate α and the death rate are presumed to vary annually. Here I allowed α to vary between 0.7 and 1.1, allowed d to vary between 0.4 and 0.7, and set β so that the deterministric equilibrium size was 100. I ran this 25 times, with each trajectory shown as a line in figure 5.7, starting the population at 50% of the deterministic equilibrium. The population model does what you might expect: on average, the population increases to levels near the deterministic equilibrium but fluctuates around that equilibrium value as the birth and death rates vary (figure 5.7). Note, however, that the population frequently drops below the starting value (50), despite the fact that the deterministic model wants to double in value during the first thirty years, and that the spread in population sizes is enormous, spanning <40 to 150.

How have researchers used these stochastic models? Many ways of course, including calculating extinction risk, as we discussed before. Unlike the density-independent case, the question of extinction risk in density-dependent models is usually not a question of "will" the population go extinct but rather "when will it happen." That is, most stochastic models with density-dependent constraints on population growth will go extinct eventually (Nåsell 2001; Norden 1982), so a lot of work examines the expected "time to extinction," given a particular model structure.

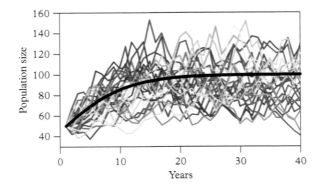

Figure 5.7 *Results of twenty-five simulations of the stochastic Beverton-Holt model. The black line shows the deterministic trajectory, while each colored line represents a unique model run.*

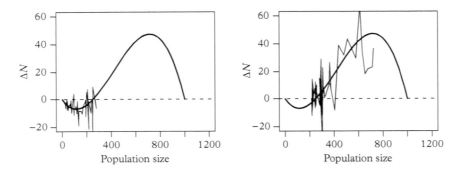

Figure 5.8 *(Left) Stochasticity doomed this population that started above the Allee threshold. The thick line is the deterministic rate of change, and the thin line is the actual rate of change. (Right) Stochasticity saved this population that started below the Allee threshold.*

These models have also been used to evaluate the interactive effects of density-dependent feedbacks (which, as you remember, can produce rather complex behavior on their own) and external stochastic drivers, and to identify the conditions under which population processes dampen these external effects (Sinclair and Pech 1996).

5.7.1 Allee effects

The Allee models are particularly interesting in stochastic models. Recall that, in our deterministic case, populations below the Allee threshold will always decrease and populations above the threshold will always grow to carrying capacity. These rules are not true in stochastic models. A population that is close to but above the Allee threshold can dip below the Allee threshold, owing to unfortunate stochastic effects (figure 7.8).

This means that a population that would be safe in the deterministic case is now at risk. At the same time, a population that is close to but below the Allee threshold can jump above the threshold if stochastic effects are in their favor (figure 5.8). In this situation, stochasticity can save a population that would otherwise certainly decline

5.8 Structured stochastic models

The killer whale extinction risk calculation might have seemed somewhat incomplete to you. These are long-lived animals that do not become sexually mature until at least a decade, females become reproductively senescent at old ages, and there are distinct differences in demographic rates for each gender. Surely, a structured population model can better account for these components of reality.

If you thought that, you are correct. The models used in decision-making are often more detailed by including population structure (Brook et al. 2000). In fact, one key issue facing the current Southern Resident killer whales is the lack of reproductively active females in the population, together with the fact that many of the most recent births have been males (Ward et al. 2013).

There is a rich field of study that applies similar methods to structured population models (chapter 3). The main distinction is that our model is modified to include stochasticity in the following way:

$$\mathbf{N}_{t+1} = \mathbf{A}_t \mathbf{N}_t \tag{5.18}$$

where now the matrix $\mathbf{A}$ has been replaced with $\mathbf{A}_t$ to denote that its elements change through time. The implementation is a bit messier but conceptually the same: you define a reasonable way to describe how each parameter value varies annually (or make the outcome subject to demographic stochasticity), update the matrix $\mathbf{A}$ each year, and then simulate the population. There are analytic solutions that tell you extinction risk that hold under a lot of conditions (Holmes 2004). One thing to keep in mind is that the starting vector of abundances can be really important. Given the same overall population abundance, the population risk is different if all of those individuals are in early life-history stages, mature stages, or post-mature stages.

We also often combine density dependence and population structure in stochastic models. For example, many models used in fisheries management for population assessment are stochastic age-structured models, where production of offspring is density dependent but also subject to environmental variability, which can be extensive in many marine fishes. For other population stages, demographic parameters are often assumed to be constant through time, although they vary with age.

A common model structure might look like this:

$$N_{a,t} = \begin{cases} f(SB_t)v_t & \text{if } a = 1 \\ N_{a-1,t-1}S_{a-1} & \text{otherwise} \end{cases} \tag{5.19}$$

where SB_t is the spawning biomass (total weight of all reproductively mature fish) in year t, $f(SB_t)$ is some density-dependent function, v_t is a random number to represent environmental influences on reproduction, and S is annual survivorship, which depends on age, fishing intensity, and susceptibility of individuals to the fishing gear.

Operationally, there will almost never be a clean, elegant solution for extinction risk for a model like this. But that's fine, because we can always just simulate many iterations of the model, and we are building pretty good intuition about how these models operate to be able to explain the simulation results.

Summary

- We add stochasticity to account for environmental processes not explicitly part of our model world.
- The arithmetic mean of λ is misleading (it tends to overestimate true population growth), but we like to use arithmetic means because they are more intuitive.
- Working in log-space resolves this problem; we can look at the arithmetic mean of the log (λ) to get a clue about the long-term behavior of the population.
- We can estimate extinction risk, even for populations that appear to be growing.
- Estimates of extinction risk are sensitive to whether autocorrelation in the λ are considered, and whether uncertainty in the mean and variance of λ are acknowledged.

Exercises

1. Use spreadsheets or R to create a density-independent model that includes stochastic variation to project a population twenty years into the future. Assume a starting population size of 100, that the mean r_t is 0, and that the standard deviation in r_t is 0.2. To do this, create a unique draw from a normal distribution to get the r_t for each year. Given the r_t, generate N_{t+1} using

$$N_{t+1} = N_t e^{r_t}$$

2. Using either R or a spreadsheet with a macro, write a script to repeat the model 1,000 times, and calculate the proportion of times the population dropped below twenty individuals.
3. Add autocorrelation to your model, assuming that ρ equals 0.25, using equation 5.14 to calculate r_t. How much did it change your estimate of the extinction risk?
4. Advanced: Create an R script or Excel spreadsheet (and macro) that will calculate extinction risk as in exercise 2, but admitting that the mean of r_t is known imprecisely, with a standard deviation of 0.05 (i.e., the $\bar{r}$ is assumed to be normally distributed with a mean of 0 and a standard deviation of 0.05).

Part II

Fitting Models to Data

6

Why Fit Models to Data?

Part 1 illustrated how we use models to make inferences about the real world. We saw how they help clarify ecological understanding, aid decision-making, and evaluate risk. In all of these cases, a model was put forward and its properties were explored so that we could draw some insight regarding the real world.

But what if instead of a model you have data? And if you have data, you likely have some hypotheses that you want to test. If you view the models that are presented so far as hypotheses written in mathematical form, then we can use statistical methods to determine which hypotheses have the greatest support.

Before going too much further, it is worth contrasting the approach presented here with the statistics that you've probably learned in your coursework. Most likely, you've been taught that inference is based on null hypothesis testing. You have a hypothesis of interest, you define a null hypothesis, and then you apply a test statistic and calculate the probability of getting a test statistic at least as extreme as your observed test statistic if the null hypothesis were true. If this probability is low, we conclude the data support the alternative hypothesis (or, rather, we reject the null hypothesis in favor of the alternative hypothesis).

Null hypothesis testing has its proponents and its critics. I won't dig too deeply here on the differences of opinions but direct the reader to relevant citations in the box below (see p. X). I will, however, highlight two important things. First, the advancement of computer power at the end of the twentieth century made it much easier to apply statistical tests that are different from the ones you learned in your introductory statistics courses. Prior to then, most model fitting and model testing had to be done by casting our ecological models as statistical models. For example, ordinary least squares linear regression (fitting linear models to data) has simple solutions that can be calculated directly from the independent and dependent variables. We therefore took our models of interest, changed them so that they looked like a statistical model (like a regression model), and then ran standard null hypothesis tests. For example, people often used the Ricker function (section 2.4.2) to relate offspring production to the density of reproductive adults, because the log ratio of offspring to adult is linear in this model. Today, our computers can easily fit all manner of models even when there aren't simple solutions (when the solution needs to be found numerically). This means we can estimate parameters for any model we wish.

Introduction to Quantitative Ecology: Mathematical and Statistical Modelling for Beginners. Timothy E. Essington,
Oxford University Press. © Timothy E. Essington 2021.
DOI: 10.1093/oso/9780192843470.003.0006

Going down the rabbit hole of statistical philosophy

People hold strong beliefs about statistical approaches for drawing inference from data. Some clarification on these might be useful. Historically, there have been two main statistical camps: frequentists and Bayesians. Frequentists' methods focus on frequency of events (outcomes) under alternative hypotheses. The idea is that you can consider the distribution of outcomes under some hypothesis if you repeated an experiment or study over and over again and then ask whether your outcome was particularly unusual in that context, especially when judged against some plausible alternative hypothesis. Bayesians, on the other hand, define probability in a completely different way and take great delight in teasing frequentists over the linguistic hurdles they must jump to explain a confidence interval.

A third statistical approach, "information theory," is less easy to categorize. Like frequentist statistics, information theory relies on the idea that you imagine a world where there are several hypothetical realizations of experimental outcomes. But it doesn't look at the distribution of those outcomes in the same way to calculate a p-value.

Below are some potentially useful readings if you wish to dip your toe into the rabbit hole:

- null hypothesis testing and relationships of different methods: Murtaugh 2014; Stephens et al. 2005; Stephens et al. 2007
- information theory: Burnham and Anderson 2014; Burnham and Anderson 1998; D. Anderson et al. 2000
- Bayesian: Clark 2004; Dennis 1996; Hobbs and Hooten 2015

Second, there are real practical limits to null hypothesis testing for many issues that ecologists and decision-makers face. The biggest is that null hypothesis testing only allows you to test one model (the one we care about) against one other model (the null hypothesis, the one that you don't care about or seek to reject). This works well in Platt's "strong inference" view of science, where excitable laboratory scientists create beautiful logical trees, each branch has two forks, and an experiment can quickly be run to determine unequivocally which fork is correct (Platt 1964). Ecological systems are far more complex: causal relationships are contextual, and ecological dynamics are often determined by multiple interacting factors (Pickett et al. 1994). Null hypothesis testing is not well suited to these types of problems. For that reason, we generally work using Chamberlain's ([1890] 1965) method of multiple working hypotheses.

The other main limit is more philosophical: if one uses an arbitrary α level to decide whether to reject or accept the null hypothesis, one is making a binary decision on a measure of support that is continuous. Why should a p-value of 0.051 lead to a radically different conclusion than a p-value of 0.049? Part of the problem is in the use of null hypothesis testing, rather than the underlying method itself. Still, perhaps it would be better if we presented all of our working hypotheses and then transparently reported the degree of support for each (Johnson and Omland 2004).

Finally, when it comes to models to aid in decision-making, standard null hypothesis tests that either accept or reject null hypotheses can often be unhelpful (Hilborn and Mangel 1997). In most cases, there are multiple potential causes for an ecological, conservation, or natural resource problem that decision-makers hope to address by enacting policies. The most appropriate and effective policy actions depend quite strongly on which of these causes are most important. Minimally, decision-makers would benefit from knowing to what extent different hypotheses can be distinguished from the available data, and whether some hypotheses can safely be dismissed from further consideration. Knowing a p-value against a null hypothesis doesn't provide this important information.

As an example, I was recently part of a scientific review panel tasked with reviewing root causes of the sharp decline of Pacific herring (*Clupea pallassi*) in Puget Sound, WA. These fish play vital roles in the food web, transferring energy from the base to the top of the food web, thereby supporting endangered Pacific salmon stocks and the killer whales that feed on salmon. Over the course of a long day, we heard compelling evidence for multiple hypotheses, including loss of spawning habitat, ultraviolet damage to eggs, toxic contaminants in the runoff from the watershed or from pilings used to construct piers, and multiple disease outbreaks. Each scientist capably reported on the evidence for "their" hypothesis. But not a single one asked, "How much of the herring decline can we attribute to any single cause?", let alone "What are the cumulative effects of these different stressors?" Ecological models, combined with data and the statistical tools that follow, will allow you to answer these types of questions.

This part of the book will focus on developing these statistical tools so that you can express hypotheses as mathematical models, fit the models to data, and assess the degree of support for each. First we'll review probability theory and common probability density functions that arise in ecological modeling. Second, we'll address model fitting, focusing on the concept of likelihood and its use for parameter estimation. Third, we'll use likelihood as the basis to judge the degree of support of alternative models. Fourth, we'll briefly explore a completely different statistical philosophy, Bayesian statistics.

As in Part 1, this part of the book will focus on concepts, while the skills section will illustrate the application of these methods in spreadsheets and in R.

7

Random Variables and Probability

7.1 Introduction

Random variables are the core concept in understanding probability, parameter estimation, and model selection. It is a term that you may learned before, but you may not have fully understood its meaning. Here we'll review the basic concept and identify the two main kinds of random variables.

7.1.1 What is a random variable?

Suppose you have a pair of dice. You shake them up, throw them on the table, and add up the two numbers. The outcome is a *random variable*. It takes on different values based on a set of probabilities that are defined for each value. A more formal definition is, any outcome that follows from a given trial, experiment, or sample event. It doesn't matter if it is a lab experiment or a field experiment. In other words, all of that beautiful data you collected? From now on, think of them as a beautiful collection of random variables.

There are two basic categories of random variables. Discrete random variables are those whose outcomes can only take discrete values. The above example of throwing two dice is a good example: there are eleven possible outcomes: all of the integers between 2 and 12. There is no way to throw the dice and get a 8.467. Continuous random variables are those whose outcomes can take any value. For example, suppose while you where throwing the dice, someone was timing how long it took for the dice to stop moving once they left your hand. This could be anything: 2.34524 seconds, 4.02042 seconds, or whatever.

Put another way, anything that you count is a discrete variable. Anything that you measure is a continuous variable.

Below we will briefly review random variables, because they form the basis of all statistics. If you feel that you have a good grasp of these ideas, skip ahead to the next section. In my experience, knowledge of statistics, probability, and random variables is lot like helium in a balloon. While you're taking a course, the balloon is full and buoyant. Come back a year later and that balloon is half-full and lying on the ground. This information just leaks out of our heads for some reason.

Introduction to Quantitative Ecology: Mathematical and Statistical Modelling for Beginners. Timothy E. Essington, Oxford University Press. © Timothy E. Essington 2021.
DOI: 10.1093/oso/9780192843470.003.0007

Properties of random variables

If a random variable, X, is discrete (e.g., the number of observed predation events), then we define the probability that an observation has some quality A as

$$P(X \text{ is } A) = \sum_{x_i \in A} P(x_i)$$

The right-hand side of this equation is the *Probability Mass Function*. It tells you the probability of each possible outcome. Say we wanted to know the probability of the dice above coming up either 5 or 6; this would simply be the sum of the probability of getting a 5 ($P(x_i = 5)$) plus the probability of getting a 6 ($P(x_i = 6)$).

If X is a continuous random variable (e.g., time until the dice stop moving), then

$$P(X \text{ is } A) = \int_{X \in A} f(x)dx$$

where the left-hand side of the equation denotes the probability that X lies in the region A (defined by lower and upper bounds of x), and $f(x)$ is the *Probability Density Function*. Note that there is an integral (instead of a summation) because x is a continuous variable.

The expected value of the discrete random variable X that can take values $x_1, x_2, x_3, \ldots, x_n$ is:

$$\mathbb{E}[X] = \sum_{i}^{n} x_i P(x_i)$$

where x_i are all of the outcomes that X can take.

The expected value of a continuous random variable X that can take any value x is

$$\mathbb{E}[X] = \int x f(x)dx$$

Random variables have various "moments," which are defined where the jth moment is $\mathbb{E}[X^j]$. The mean of X is the first moment and equals $\mathbb{E}[X]$ (e.g., the expected value of X). The second moment is similar, but, rather than find $\mathbb{E}[X^2]$, we instead look at the "centered" moments (meaning that they are centered around the mean). Thus, the second moment is $\mathbb{E}[(X - \mu)^2]$, where μ is the mean. In other words, the second moment tells us the expected squared deviation from the mean. This is called the variance, and the square root of the variance is called the standard deviation.

7.2   Binomial

The binomial probability mass function describes the number of "successes" in a series of discrete trials, where all trials have the same probability of success. The simplest example of a binomial probability mass function is a series of coin tosses, where a success means the coin comes up heads. If you were to flip a coin ten times, and each time there is a 50% probability of the coin coming up heads, you would use the binomial probability density function to estimate the probability of getting zero heads, one heads, two heads, and so on.

7.2.1   Key things about this distribution

The outcome is discrete (you can never have 4.78 heads) and derives from a series of discrete trials (the coin tosses, which are also discrete, as you can't flip a coin 4.78 times). It assumes that trials are independent and have identical properties (i.e., the probability of getting "heads" is the same for each toss).

7.2.2   The probability mass function

We denote the outcome as X. Here, X refers to the number of times the coin turns up heads, N is the number of trials (the number of coin tosses), and p is the probability that any given coin toss comes up heads. From this, we can calculate the probability of all outcomes between $X = 0$ and $X = N$:

$$P(X = x) = \frac{N!}{x!(N-x)!}p^x(1-p)^{N-x} \tag{7.1}$$

where the notation $N!$ and $x!$ refer to N factorial x factorial, respectively. The left-hand side says that X is a random variable that can take any value x, while the right-hand side tells us the probability that it will take any value x.

7.2.3   When would I use this?

The coin toss is the easiest way to think about this, but it applies to any analogous situation where you are counting the number of times a given event happens over several discrete trials, where there is a constant probability of the event happening for each trial. So, for instance, the probability of death over survival, a prey encounter over no prey encounter, and successful seed germination over unsuccessful germination are all discrete outcomes that happen over a series of discrete trials.

7.2.4   Properties of the function

The expected value of the outcome is $\mathbb{E}[X] = pN$, which makes sense because it is saying that if the probability of a coin toss heads is p, and you flip the coin N times, on average you will get p times N heads.

The variance of the binomial equals $Np(1-p)$. This means the variance grows as N gets larger. This also makes some sense—there simply aren't that many different outcomes if you only flip the coin once or twice, but there are many different outcomes that can happen if you flip the coin 100 times. The variance is also determined by the product $p(1-p)$. If p is close to 0 or close to 1, this product will be fairly small. Think about why the variance of outcomes is small if p is very nearly 0 or very nearly 1.

7.2.5   Example

Suppose you have a stage-structured population, and you have twenty individuals in a reproductively immature stage. You wish to project the number of individuals that transition into a mature stage over some discrete time interval. The number of individuals that mature is a random variable described by a binomial probability distribution, because the randomness acts on discrete individuals. If the probability is low, $p = 0.1$ (which might happen for small or young individuals), most of the individuals will not become mature, but you may reasonably get as many as five or six that become mature (figure 7.1) over some time increment. If, instead, the maturation probability is $p = 0.5$, then there is a much wider range of outcomes (figure 7.1), consistent with the finding that the variance is proportional to p times $(1-p)$.

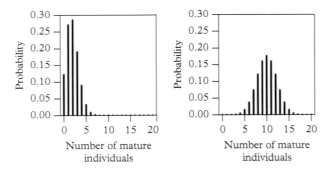

Figure 7.1 *The probability of seeing* x *individuals become mature out of 20 individuals, when the probability of maturation is p = 0.1 (left), and p = 0.5 (right).*

7.3 Poisson

The Poisson probability mass function describes the number of "successes" over a fixed, continuous sampling frame where, at any moment in time or space, there is a some rate of success. The number of "successes" might be the number of grizzly bears (*Ursus arctos horribilis*) observed using an aerial survey. Unlike the case for the binomial probability mass function, the observations come from looking over a continuous sampling frame, for example, a certain amount of area that is surveyed. Alternatively, one could monitor one location for a set period of time and count the number of grizzly bears that are observed. The probability of seeing a grizzly bear at any moment is related to the density of grizzly bears.

7.3.1 Key things about this distribution

The key things about this distribution are that the outcome is discrete (you can never see 4.78 bears perhaps) and that it derives from a continuous observation window, over either time or space. Also, it presumes that the probability of an outcome (or, rather, the event density) is constant over space or time.

7.3.2 The probability function

We denote the outcome as X. Here, X refers to the number bears that are observed, t is the area or amount of time that is surveyed, and r is the density of grizzly bears. From this, we can calculate the probability of all outcomes for $X \geq 0$:

$$P(X = x) = \frac{e^{-rt}(rt)^x}{x!} \tag{7.2}$$

Notice that r and t are always multiplied by each other. Sometimes you will see the model written like this:

$$P(X = x) = \frac{e^{-\lambda}(\lambda)^x}{x!} \tag{7.3}$$

Here r times t has been replaced by λ. You would use this if you cared about the total numbers of bears present in a defined space, rather than the density.

Remember that the left-hand side says that X is a random variable, and we can calculate the probability that X takes any value x, using the right-hand side. The example of surveys over space is the easiest way to think about this, but it applies to any analogous situation where you are counting the number of times an event happens over a continuous sampling frame and there is a constant probability of the event happening.

7.3.3   When would I use this?

You would use this whenever your observations are discrete counts of events over a continuous time or spatial observation window. Most ecological surveys, for instance, have a defined space. You might place a quadrat on a forest floor and count the number of seedlings in the quadrat. The number of seedlings could be described by a Poisson. Needless to say, this is a very common probability density function in ecological research.

7.3.4   Properties of the function

The expected value of the outcome is $\mathbb{E}[X] = rt = \lambda$, which makes sense because it is saying that if the rate of a seeing a "success" is r per given area or time, and you search over area $= t$ or time $= t$, then, on average, you will see r times t "successes." If you think of r as a density, and the true grizzly bear density is 3 bears/km^2, and you search 10 km^2, you expect, on average, to see 30 bears. The variance of the Poisson also equals rt. This is a bit less intuitive, but it means that the ratio of mean to variance always equals 1. This is quite different from many other probability density functions, where the mean and variance are uncoupled.

7.3.5   Example

You have a camera attached to a foraging emperor penguin to examine its foraging behavior. From the video feed, you have continuous-time monitoring of the number of attacks on fish prey. If the Poisson rate parameter is 0.1 min^{-1} and you have fifty minutes of observation, then you might reasonably expect to see from zero to about twelve fish attacks (figure 7.2), although in rare cases you might see many more than twelve. If the Poisson rate parameter is 0.3 min^{-1}, then there is a much wider range of likely outcomes, ranging from five to twenty-five (figure 7.2). This is consistent with our expectation that the variance increase (in this case) is equal to the product r times t.

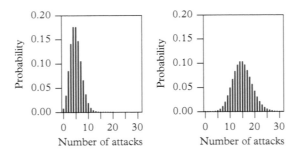

Figure 7.2 *Probability of observed attacks by penguins when r = 0.1 (left) and r = 0.3 (right).*

7.4 Negative binomial

The negative binomial is used in contexts that are similar to those in which the Poisson would be used: it describes the number of "successes" over an observation window that is continuous in time or space, where, at any moment in time or space, there is a some probability of success. The difference is that there is an additional parameter that describes patchiness in success over time or space.

Another way to think about this is that the grizzly bear density (the parameter r) in the example given above varies over the sampling frame. Thus, in some areas, the density of bears is very high; in others, it is very low. This makes the negative binomial more flexible than the Poisson, because the variance and the mean are no longer required to be equal (the variance will be equal to, or exceed the mean).

7.4.1 Key things about this function

The outcome is discrete and derives from a continuous observation window, over either time or space. Unlike the Poisson, it does not presume that the rate of seeing a grizzly bear at any given location is constant. We tend to use it when we think the event rate r varies across the sampling frame.

7.4.2 When would I use this?

You would use this in most of the same contexts for which you would use the Poisson, especially when the the variance is greater than the expected value. That is, for very patchy encounters, the negative binomial is a good alternative

7.4.3 The probability function

The negative binomial is the source of a great deal of confusion, because it can be derived in different ways and can be expressed using different equations. The classic derivation, and the one that explains its name, is the expected number of failures in a series of discrete trials, before a success happens. It is therefore called a "negative binomial" in a nod to its roots in the binomial distribution, and owing to the fact that it describes the counts of failures (the "negative" part of the name). In ecological circles, the derivation is different. Instead, it starts with the Poisson distribution but assumes that r varies over space and time, with some mean value. The random variable is X, which can take any discrete outcome x. In our example, X refers to the number of bears that are observed, t is the area or amount of time that is surveyed, and r is the rate of observing grizzly bears over some units of time or space. The parameter k has no immediate ecological meaning, but it sets the level of patchiness (variance) of the grizzly bear encounters. Small values of k produce very patchy (high-variance) encounters. From this, we can calculate the probability of all outcomes for $x \geq 0$:

$$P(X = x) = \frac{\Gamma(x+k)}{\Gamma(k)x!} \left(\frac{rt}{k+rt} \right)^x \left(\frac{k}{k+rt} \right)^k \tag{7.4}$$

The notation $\Gamma()$ refers to the gamma function (you can think of the gamma function in the same way you think of logarithms—it is a specific mathematical operation, but the specific details of this operation are not critical to know at this point). In Excel, you can get the natural log of the gamma function using the `gammaln()` command, while in R you can get the gamma function calculated at any number by using the `gamma()` command.

7.4.4   Properties of the function

Just like the case for the Poisson distribution, the expected value of X is $\mathbf{E}[X] = rt$. However, the variance of the negative binomial now equals $rt + (rt)^2/k$. In other words, the variance becomes inflated relative to the Poisson by an additional amount equal to $(rt)^2/k$. Small values of k produce high variance.

7.4.5   Example

In our Poisson example, we considered the case of counts of prey encounters from continuous monitoring of emperor penguins. That case assumed that a penguin was equally likely to encounter a prey at any moment in time. There is one huge problem with that assumption: penguins live on ice and dive into water to find fish prey. While they are on ice, they have zero chance of finding food. Although not a perfect example of the negative binomial, it illustrates the idea that the event probability varies over the sampling frame.

When k is large, say, $k = 25$, the expected number of prey attacks looks fairly similar to that in the Poisson model. Compare the left panel of figure 7.3 to the right panel

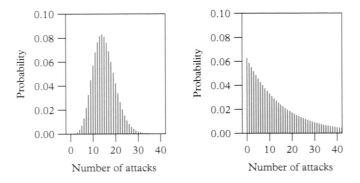

Figure 7.3 *Two examples of the negative binomial probability function, each with the same expected value but with different variances. (Left) $k = 25$; (right) $k = 1$. Note that, on the right panel, the right-hand tail of the distribution extends beyond the plot range.*

of figure 7.2. When k is small, say, $k = 1$, the probability density is markedly different (figure 7.3). Notice now that the most common outcome is zero prey encounters. At the same time, it would not be too strange if there were forty or more prey encounters.

7.5 Normal

The normal distribution describes the outcomes of events that are continuous variables. For example, temperature can take any value, not just integer values. The distribution of temperatures is described by any one of several continuous probability density functions. The normal probability density function is a common choice, because it has many desirable features. First, the mean and the variance are defined directly based on its two parameters. Two, it has many helpful statistical properties, such as that given by the central limit theorem, which states that if you have a big enough sample size, the distribution of the mean of the samples is approximately normally distributed.

7.5.1 Key things about this distribution

The outcome is continuous, which means that the probability that a random variable takes any particular value is infinitesimally small. We often bin our observations (e.g., what is the probability that the temperature will be between 12 and 13 degrees) and then integrate to get the probability.

7.5.2 When would I use this?

Whenever you can get away with it. The normal distribution has so many useful features that you might even choose to use it when your observations are discrete (typically, you would do this when the discrete observations are large numbers, say, greater than 100).

7.5.3 The probability density function

Because the normal distribution is continuous, we use a probability density function, which itself has to be integrated to get the probability that a random variable is between two numbers. The normal distribution has two parameters, μ and σ. The probability density is

$$f(x) = \frac{1}{\sqrt{2\pi\sigma^2}} \exp\left(-\frac{(\mu - x)^2}{2\sigma^2}\right) \tag{7.5}$$

where here I adopt the notation $\exp(y)$ to refer to e raised to the power y, which makes the equation easier to read. To get the probability of some set of possible outcomes, k, we integrate over the range of values of k that we are interested in. Assume our range of k is defined by a lower bound, k_{lower} and upper bound, k_{upper}; then

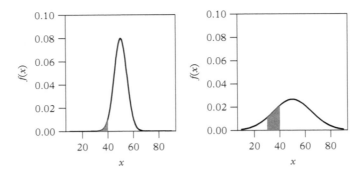

Figure 7.4 *The probability of events are calculated by integrating to get the area under the curve. Here, the probability that the tree diameter is between 30 and 40 cm, given a mean of 50 cm and a standard deviation of 5 (left) and 15 (right).*

$$P(k_{\text{lower}} < x < k_{\text{upper}}) = \int_{k_{\text{lower}}}^{k_{\text{upper}}} \frac{1}{\sqrt{2\pi\sigma^2}} \exp\left(-\frac{(\mu - x)^2}{2\sigma^2}\right) dx \qquad (7.6)$$

7.5.4   Properties of the function

One reason why the normal distribution is so useful is that the mean and the variance are simply the two parameters of the distribution, that is, $\mathbb{E}[X] = \mu$, $\mathbb{E}[(X - \mu)^2] = \sigma^2$.

7.5.5   Example

Suppose you are measuring tree diameter in a forest. The average tree diameter is 50 cm and has a standard deviation of 5 cm. You want to know the probability that any given tree will have a tree diameter between 30 and 40 cm. If the standard deviation is small, the probability will be relatively low, less than 3% (figure 7.4). If the standard deviation is larger (15), the area under the curve between 30 and 40 is much larger, 16%.

7.6   Log-normal

The log-normal distribution describes the outcomes of events that are continuous variables, where the logarithm of the observations are themselves normally distributed. The log-normal distribution structure is fairly common in ecological data, and there are theoretical reasons for why it arises in nature (Hilborn and Walters 1992). Practically speaking, it is often convenient because the outcome is always a positive number, which means we can use it when we want to ensure that our random variable exceeds 0.

7.6.1   Key things about this distribution

Because the log-normal is the distribution of a variable whose logarithm is normally distributed, the variable cannot be 0 or lower. In other words, the log-normal distribution is only defined for positive values.

7.6.2   When would I use this?

The log-normal is frequently used for continuous random variables that have to be greater than 0.

7.6.3   The probability density function

It is easiest to break the function into two components. The first is the definition of our random variable X:

$$X = e^Y \tag{7.7}$$

where Y is normally distributed with mean μ and variance σ. Given these assumptions, the probability density function of X equals

$$f(x) = \frac{1}{x\sqrt{2\pi\sigma^2}}\exp\left(\frac{(\mu - \log(x))^2}{2\sigma^2}\right) \tag{7.8}$$

7.6.4   Properties of the function

You might expect that the mean of the log-normal is simply e^μ, but it turns out that the mean of the log-normal is $\mathbb{E}[X] = e^{\mu + 0.5\sigma^2}$. This may look vaguely familiar to you (section 5.3). If you want to generate log-normal errors with a specified mean, it's important to correct for the variance. The coefficient of variation (standard deviation divided by the mean) is approximately equal to σ.

7.6.5   Example

We have already seen examples of the log-normal distribution when exploring stochastic density-independent models. Recall that we defined r_t as the log of the λ_t and that r_t were random draws from a normal distribution with mean μ and standard deviation σ. In other words, λ_t equals $\exp(r_t)$. Thus, λ_t, the annual population growth rate, is a log-normal random variable with mean equal to $e^{\mu + \sigma^2/2}$. In addition, the population size at any time t steps into the future is a log-normal random variable with mean equal to $N_0 e^{(\mu + \sigma^2/2t)}$.

7.7   Advanced: Other distributions

7.7.1   The gamma distribution

Like the log-normal distribution, the gamma is only defined for values of X that exceed 0. However, it can take different shapes, which sometimes is useful in fitting models to data (figure 7.5). The probability density function is

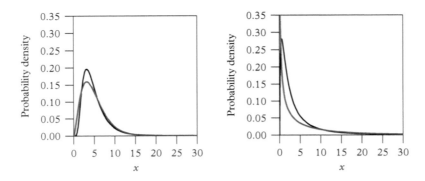

Figure 7.5 *Comparison of the log-normal (black) and the gamma distribution (blue). All distributions have a mean of 5; distributions on the left have a standard deviation of 3, and distributions on the right have a standard deviation of 9.*

$$f(x) = \frac{\beta^{\alpha}}{\Gamma(\alpha)} x^{\alpha-1} e^{-\beta x} \tag{7.9}$$

The mean is α/β and the variance is α/β^2.

7.7.2   The beta distribution

The beta distribution is a common distribution to use when modeling random variables that must take values between 0 and 1. For instance, in a structured stochastic population model, where you want the survival or transition probabilities to be random draws, the beta distribution is ideal. The probability density function is

$$f(x) = \frac{\Gamma(\alpha+\beta)}{\Gamma(\alpha)\Gamma(\beta)} (1-x)^{\beta-1} x^{\alpha-1} \tag{7.10}$$

The mean of the beta distribution is $\frac{\alpha}{\alpha+\beta}$. The function is not defined for $x = 0$ or $x = 1$.

7.7.3   Student's t-distribution

Student's t-distribution is an increasingly valued substitute for the normal distribution (Anderson et al. 2017). The normal distribution has one undesirable feature: as you move out on the tails, the probability of events decline exponentially (Taleb 2010). This means that extreme events are quite unlikely. The Student's t-distribution allows for greater probability of events that are far away for the mean, even with the same overall variance as the normal distribution. The probability density function is

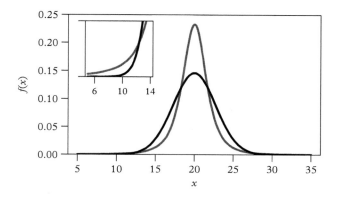

Figure 7.6 *Student's t-distribution (gray) and normal distribution (black). Both distributions have the same mean and variance. The key difference lies in the tails of the distributions. Inset shows the difference in probability density for small values of x.*

$$f(x) = \frac{\Gamma\left(\frac{v+1}{2}\right)}{\Gamma\left(\frac{v}{2}\right)\sqrt{\pi v \sigma^2}} \left(1 + \frac{1}{v}\frac{(x-\mu)^2}{\sigma^2}\right)^{-\frac{v+1}{2}} \tag{7.11}$$

where v is the degrees of freedom ($v \geq 1$), the mean of the distribution is μ, and the variance is $\frac{v}{v-2}\sigma^2$ (for $v \geq 2$). Note here that the term σ acts as a scaling parameter so that you can make the Student's t-distribution have whatever overall variance you wish. For comparison, consider a normal distribution with mean 20 and variance 7.5. The probability of any event less than 10 is vanishing small (figure 7.6). Yet, a Student's t-distribution, with same mean and variance (here with $v = 3$), can allow events very far from the mean to occur more frequently (figure 7.6).

7.7.4   The beta-binomial distribution

The beta-binomial pairs with the binomial distribution in the same way that the negative binomial pairs with the Poisson distribution. Recall that we use the negative binomial with discrete count data when the data are more dispersed (there is more variability) than the Poisson model can account for. The beta-binomial does the same thing for the binomial distribution. It presumes that the probability of some outcome over some series of discrete trials is not the same over all trials, but instead varies according to a beta distribution. Given this assumption, the probability mass function of the beta-binomial distribution is

$$P(X = x) = \frac{N!}{x!(N-x)!} \cdot \frac{\Gamma(\theta)}{\Gamma(p\theta)\Gamma((1-p)\theta)} \cdot \frac{a\Gamma(x+p\theta)\Gamma(N-x+(1-p)\theta)}{\Gamma(N+\theta)} \tag{7.12}$$

where p is the average probability of the outcome happening in a trial, and θ is the overdispersion parameter that governs the variance in p. The expected value is the same as the binomial (pN), but the variance is given by $Np(1-p)\left(1 + \frac{N-1}{\theta+1}\right)$.

7.7.5   Zero-inflated models

Sometimes ecological count data have a lot of zeros, and these cannot be accounted for by our standard probability mass function such as the Poisson or even the negative binomial. In these cases, we might apply a so-called zero-inflated or mixture model. It's called a mixture model because it combines both the binomial model and another model. Usually, we say that the true density is 0 with some probability p and has a nonzero density at probability $1-p$. For instance, a zero-inflated Poisson model of count data is

$$P(X = x) = \begin{cases} p + (1-p)e^{-\lambda} & \text{if } x = 0 \\ (1-p)\frac{e^{-\lambda}\lambda^x}{x!} & \text{otherwise} \end{cases} \qquad (7.13)$$

where p is the binomial probability per trial, and λ is the mean of the Poisson. Notice in the first line that there are two ways that we could get a 0. One is that the thing we are counting is simply not there with probability p. The second is that the thing we are counting could be present (with probability $1-p$), but we just didn't see them as given by the Poisson (i.e., $e^{-\lambda}$).

Summary

- All data are random variables; that means data can be explained as coming from probability mass functions or probability density functions.
- There are many different probability functions; you choose the function based on some knowledge of the process that generated the random variable.
- You can calculate probabilities of future events easily once you've chosen your probability density function and specified its parameters.
- For continuous probability density functions, you need to integrate the function to get the probability that the random variable takes a value within any specified range.

Exercises

1. You are going to run an experiment to test the effects of a toxic contaminant on juvenile salmon mortality rates. You will estimate mortality rate as the fraction of salmon that died in the experiment. You choose to run your experiment with twenty salmon. What do the parameters N, p, and k in the binomial probability density function represent in this experiment?

2. If the true mortality rate is 0.5, what is the probability that the fraction of salmon that die in the experiment is within 0.1 unit of the true mortality rate?

3. Repeat the calculation in exercise 2 but with true mortality rates of 0.25 and 0.75. Describe and explain the differences (and similarities) in your probability· calculation for the three different mortality rates.

4. Geoduck clams are among the most valuable shellfish in the world. They live in nearshore habitats and can be surveyed by divers monitoring the seabed. Suppose the true density is 0.1 geoducks/m^2. You wish to know the probability of seeing a specified number of geoducks, given some sampling area.

   Why is the Poisson an appropriate probability density function? What do the parameters r and t represent in this situation?

5. Given the assumption that geoduck densities are described with a Poisson probability mass function, with r equal to 0.1/m^2, what is the minimum sample area needed so that there is an 80% or greater chance that the observed number of geoducks per square meter is within 0.05 of the true density?

8

Likelihood and Its Applications

8.1 Introduction

Likelihood is the link between data and models. We use likelihood to estimate model parameters. We use likelihood to judge the degree of precision of parameter estimates. And we use likelihood to weight support for alternative models. Likelihood is therefore a crucial concept that underlies our ability to test multiple models.

We introduce the concept of likelihood by first remembering how we applied probability distributions to calculate the probability of future events. Suppose I flipped a coin ten times. From the binomial probability density function, you can calculate the probability of several outcomes. If it is a fair coin, then $p = 0.5$, and there are eleven possible outcomes, each of which has an easily calculated probability (table 8.1).

Table 8.1 *Probabilities for each possible outcome for ten coin flips with a fair coin*

# Heads	Probability
0	0.001
1	0.01
2	0.044
3	0.117
4	0.205
5	0.246
6	0.205
7	0.117
8	0.044
9	0.01
10	0.001

Introduction to Quantitative Ecology: Mathematical and Statistical Modelling for Beginners. Timothy E. Essington, Oxford University Press. © Timothy E. Essington 2021.
DOI: 10.1093/oso/9780192843470.003.0008

Suppose I actually flipped a coin ten times and came up with eight heads. You would conclude that this is a somewhat unlikely but not totally implausible outcome (it should happen about 4% of the time).

8.1.1   Was this a fair coin?

Now suppose that there were some stakes on this series of coin flips. Let's say I made a bet with you. If I get seven or fewer heads out of ten coin tosses, then I give you $20. Otherwise, you give me $200. If I got eight heads in this coin flip, you might rightfully question whether the coin was fair. The above calculations presumed that the probability of getting heads on any given flip was 0.5, but, given the high stakes, maybe I had a trick coin.

In essence, we have reversed the line of reasoning. In the first case, you knew the parameters of the probability density function and used them to calculate the probability of future outcomes. Now, you have an outcome, and you want to make some inference about the parameters of the probability function. Specifically, you need a way to estimate the parameter p based on the outcome and to determine whether it is close to $p = 0.5$.

8.1.2   Likelihood to the rescue

Likelihood is the probability of getting some known outcome, when the underlying parameters of the probability function, those that we wish to estimate, are not known. In the case above, you want to know (i.e., estimate) the likelihood that $p = 0.5$. We calculate this, which is denoted $L(p = 0.5)$, as the probability of getting the outcome ($x = 8$), if p really were equal to 0.5:

Likelihood (parameter) = P(outcome | parameter)

We call this "likelihood" because it is measuring how likely the observed outcome is if a particular parameter or model were true. If the outcome is unlikely, then the outcome does not provide much support for that parameter value or model. If the outcome is very likely, then the outcome provides stronger support for that parameter value or model.

8.1.3   Maximum likelihood estimation

Take yourself back in time to the late 1800s and early 1900s. The fledgling field of statistics was confronting the idea of how to estimate things from data. Take something simple, like the mean of a distribution. You have a sample, and a sample mean. What is the best way to estimate the true population mean, given the data? Or, more generally, how can we estimate any set of parameters θ, given a sample?

A third-year undergraduate student named R. A. Fisher made the following proposal: "The most probable set of values for the θ's will make P a maximum, where P is the probability of the observations, given the parameter set θ." (Fisher 1912, 156).

All manner of bickering ensued, and even Fisher acknowledged that he had been somewhat sloppy in his terminology, leading many to believe that he was adopting a Bayesian interpretation of probability and inference (Edwards 1974; Aldrich 1997). Adding to the confusion was that he hadn't named his new concept. Some eight years later, he introduced the term "likelihood," which forms the basis of modern-day statistical terminology (Edwards 1974).

To put it more plainly, we wish to estimate a parameter that is unknown, based on data that are known. The concept of maximum likelihood is fairly simple. Our best estimate of a parameter value is the one that makes the outcome as likely to have occurred as possible. This sounds like a mouthful, so it's useful to think about the opposite case. Suppose I flipped a coin ten times and got eight heads. There is a 0% chance of getting that outcome if the true probability of getting a head was 1. So I would be incredibly dimwitted to use this value as my estimate of p, given the data. In fact, I would be choosing the parameter that makes the outcome as *unlikely* to have happened as possible (in this case, choosing a parameter value for which there was zero chance of getting the observed outcome). If I were smart, I'd do the opposite—I'd use the value of p that makes the outcome as likely to have occurred as possible. The data are most likely to have happened if $p = 0.8$. In other words, the maximum likelihood estimate is $p = 0.8$, given the data.

The principle of *maximum likelihood estimation* simply says that the best fitting parameter values are the ones where the outcome had the greatest chance of occurring. When we say "maximum likelihood," we are careful not to imply that the parameter value is most likely. Only that the *outcome is most likely to have occurred*, given those parameter values.

Mathematically, likelihood and probabilities are calculated in exactly the same way. The only difference is what is known and what is unknown (table 8.2).

8.1.4 What likelihood is not

We define the likelihood of a parameter value as the probability of the outcome, given the parameter value (P(outcome | parameter)). It is tempting to flip that around and interpret likelihood as the probability that a particular parameter value is true, given the data:

$$P(\text{parameter}|\text{outcome}) \neq P(\text{outcome}|\text{parameter}) \tag{8.1}$$

Table 8.2 *Distinction between probability and likelihood*

Probability	Likelihood
Outcome is unknown	Outcome is known
Parameters are known	Parameters are unknown

Here is an example to illustrate why one cannot equate conditional probabilities in this way. Consider the following two statements, and ask yourself if you expect the two to be equal:

– the probability that you drink tea, given that you are from Great Britain
– the probability that you are from Great Britain, given that you drink tea

The first probability is fairly high (at least, given all of my preconceptions about Great Britain), but the second probability is extremely small, given the enormous number of tea drinkers around the world and the comparatively small number of people that live in Great Britain.

Likelihood notation

If you spend much time reading statistical papers or books, you will likely see a few different ways to express likelihood. Here, I will use $L(\theta)$ to indicate the likelihood of some model, hypothesis. or parameter values indicated by θ, given some data, x. Many use a different notation to make the condition on the data more explicit: $L(\theta|x)$. Students often complain about this notation because it invites the misinterpretation of likelihood as the probability that θ is true. Still, this is the most common notation that you will see. You also may see the L replaced with $\mathcal{L}$, that is, $\mathcal{L}(\theta|x)$.

8.2   Parameter estimation using likelihood

Let's return to the case where you flipped a coin ten times and got eight heads. The question of interest is whether the coin was fair (whether $p = 0.5$). If I assume the random variable was generated by a binomial probability mass function, then I am implicitly assuming that there is a single, identical value of p that applies to each coin toss, that is, the data are independent and identically distributed. I want to make an inference about the value of p from the outcome. The likelihood of any given value of p equals the probability of the outcome if that value of p were true. Looking at table 8.1, the outcome (eight heads) will happen 4.4% of the time if it is a fair coin. In other words, the likelihood for $p = 0.5$ equals 0.044.

We can repeat this calculation using the binomial probability function (equation 7.1) for many different values of p, each time calculating the probability of getting eight heads. I've done that in table 8.3, for values of p ranging from 0.25 to 1.0, in increments of 0.05.

Notice that the likelihood for $p = 0.5$ (fair coin) is relatively small compared to the likelihood that $p = 0.8$. In fact, the data are most likely to occur if $p = 0.8$, which makes sense, given that we had eight heads out of ten coin tosses.

Table 8.3 *Likelihood for values of p*

p	Likelihood
0.25	0.000
0.3	0.001
0.35	0.004
0.4	0.011
0.45	0.023
0.5	0.044
0.55	0.076
0.6	0.131
0.65	0.176
0.7	0.233
0.75	0.282
0.8	0.302
0.85	0.276
0.9	0.194
0.95	0.075

There is nothing magical about likelihood or estimating parameters using maximum likelihood. Calculating likelihood is essentially the same as calculating probabilities of future outcomes. Both use probability functions. But, instead of predicting the probability of future outcomes, we calculate how likely a given outcome would be if a particular parameter value were true.

Steps in calculation

Specify known parts of the probability function, including the observation. Here, we know the outcome $x = 8$, and that $N = 10$

Specify values of the parameter for which you want to calculate the likelihood. Here, you might try p from 0.05 to 0.95 in small increments of 0.0025.

For each of the candidate parameter values, calculate the likelihood (the probability of the outcome, for each possible parameter value).

The maximum likelihood parameter value is the one that has the highest likelihood.

8.3   Uncertainty in maximum likelihood parameter estimates

In our coin-flipping example, what we really wanted to know was whether the coin flip was *not* fair, that is, whether the probability of getting a heads is really different from 0.5. We found that the maximum likelihood estimate provides the best estimate, $p = 0.8$. But this didn't answer the question of whether p is not equal to 0.5. To answer this, we need to quantify a plausible range of parameter values, given the data, as a measure of our uncertainty of our parameter estimate. If we define a plausible range, and that range includes 0.5, then the outcome does not allow us to dismiss the hypothesis that it was a fair coin. If we define a range that does not include 0.5, then you suspect more strongly that the coin was unfair. If this language sounds familiar to you, it is because it is the same line of reasoning used in null hypothesis testing.

8.3.1   Calculating confidence intervals using likelihoods

We need to define some likelihood level that we deem to be too small to give a parameter value serious consideration.

The first step is to calculate likelihood for a wide range of parameter values (figure 8.1). For demonstration purposes, I've drawn a horizontal line depicting the highest likelihood that was calculated over all of these parameter values.

Next, we need to define what we will deem to be a "small" likelihood. This will always be some fraction of the largest calculated likelihood value. If you want a 95% confidence

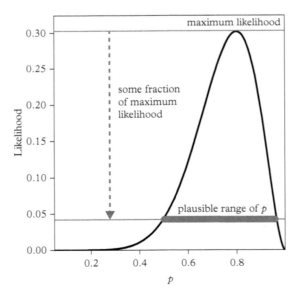

Figure 8.1 *Calculating confidence intervals from likelihood.*

interval, it should be roughly e^{-2}, or if you want to be more precise, you should use $e^{-1.92}$. We will define our range of plausible values of p as those that have likelihood equal to or greater than the lower line shown in figure 8.1.

Advanced: What fraction of maximum likelihood?

You may have noticed that the "precise" fraction to use is $e^{-1.92}$. Many students ask, "Where does that come from?" It derives from null hypothesis testing and likelihood ratios. The 95% percentile of a chi-square distribution with one degree of freedom is 3.84; one-half of this is 1.92. This is the critical value one would use in null hypothesis testing to deem one parameter worse than another when taking the difference in log-likelihoods. In other words, it gives you a 95% confidence interval.

Here, the plausible range (or confidence interval, if you prefer), ranges from 0.5007 to 0.9636.

The threshold for calculating confidence intervals is easier to identify when working in log-space. That is, you can take the logarithm of the likelihoods and then find the parameter value whose log-likelihood is 1.92 units smaller than the largest log-likelihood. For a variety of reasons, it is often easier to work with the so-called negative log-likelihood, which is simply the log-likelihood multiplied by -1. So, the best-fitting value of the parameter has the smallest negative log-likelihood, and the confidence intervals are values of the parameter producing a negative log-likelihood that is 1.92 units larger than the smallest negative log-likelihood.

To summarize, likelihood expressed as log-likelihood looks like the left panel in figure 8.2. The likelihood expressed as negative log-likelihood looks like the right panel in figure 8.2. Notice that the range of values of p are identical when seeking to maximize the log-likelihood or minimize the negative log-likelihood.

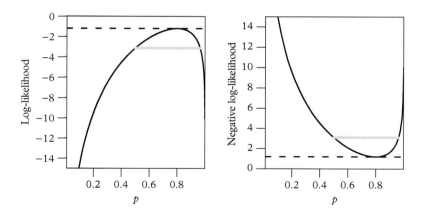

Figure 8.2 *Likelihood plotted as log-likelihood (left) and negative log-likelihood (right).*

8.3.2   To summarize

We can use likelihoods to calculate confidence intervals by

1. calculating likelihood for a range of parameter values,
2. determining the likelihood threshold, as some fraction of the largest likelihood,
3. finding the range of parameter values that have likelihoods greater than or equal to the likelihood threshold.

It is often easier to work with log-likelihoods or negative log-likelihoods.

Note that this example was relatively straightforward as we had only a single parameter to estimate. Soon we'll expand this approach to calculate confidence intervals for models with multiple parameters.

8.3.3   Practice example 1

Geoducks (pronounced "goo-ee-ducks") are one of the most valuable shellfish on earth. They are large bivalves that burrow into the sand in intertidal and subtidal habitats. Management in Washington State is conducted by auctioning harvesting rights to tracts of geoduck habitat. Before doing this, the state of Washington surveys the tract to estimate the density of geoducks via scuba-diving over several plots of geoduck habitat. Suppose you surveyed one 20 m^2 plot and counted two geoducks.

You want to know the maximum likelihood estimate for geoduck density, and a confidence interval around that, based on this single observation. The procedure is as follows:

1. Choose the correct probability function. The data are discrete, because they are counts of geoducks, and were counted over a fixed, continuous area. Because we have discrete observations from continuous observation windows, we might want to use the Poisson probability mass function.
2. Set up the range of values of the parameter to be estimated. We are interested in the geoduck density, which is equivalent to r in the Poisson probability density function. We will create a list of candidate values of r, ranging from 0.01 to 0.5 in increments of 0.0025.
3. Calculate negative log-likelihoods. For each candidate value of r, we calculate the negative log-likelihood using $t = 20$, $x = 2$ in the Poisson probability mass function (equation 7.2).
4. Identify the value of r that minimizes the negative log-likelihood. This value of r is the maximum likelihood estimate.
5. Calculate lower and upper bounds on the confidence interval. Add 1.92 to the smallest negative log-likelihood. Find the smallest value of r that has a negative log-likelihood less than or equal to this number. This is your confidence interval lower bound (note, your precision will depend on how finely you explored values of r; see the skills section in section 16.3 for more comments on this point). Now

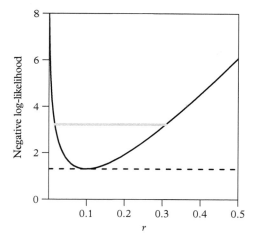

Figure 8.3 *Negative log-likelihood versus the Poisson parameter* r.

find the largest value of r that has a negative log-likelihood less than or equal to this number. This is your confidence interval upper bound.

You should get a result that looks like figure 8.3, where $r = 0.1$ is the maximum likelihood estimate and $(0.0175, 0.3075)$ is the confidence interval.

Learn how to estimate parameters and confidence intervals using more complex methods in section 16.1 (pp. 263–73).

What about continuous probability density functions?

We tackle these the same way, though we work with the probability density, $f(x)$, rather than $P(x)$ as we've done with the discrete probability mass functions. So, we'd take $\log(f(x))$ for a range of values of parameter values, find the value that maximizes the likelihood, and then calculate the confidence interval in the same way.

8.4 Likelihood with multiple observations

Suppose you continued your geoduck surveys by repeating your earlier work four times. Remember that each plot is 20 m^2. You count zero, three, six, and seven geoducks over four plots. You want to know the probability of seeing *all* four outcomes, given a particular geoduck density. The likelihood of r equals the probability of seeing zero geoducks *and* three geoducks *and* six geoducks *and* seven geoducks. When we see "and" statements in probability, we know that we have to multiply the probabilities together:

$$L(r) = P(0 \text{ geoducks} \,|\, r) \times P(3 \text{ geoducks} \,|\, r) \times P(6 \text{ geoducks} \,|\, r) \times P(7 \text{ geoducks} \,|\, r) \tag{8.2}$$

which means that, more generally, for any n observations, you would calculate the likelihood as

$$L(r) = \prod_n P(x_n|r,t) \tag{8.3}$$

We can do this for our observations above, but now is a good time to introduce a useful trick. Namely, the individual probabilities on the right-hand side are all, by definition, numbers that are between 0 and 1. If we have a very large data set, say, 1,000 measurements, the product of all of these probabilities is going to be very small. So small, in fact, that our computers might be tempted to round them down to zero. We don't want that to happen, because then we can't compare likelihoods. An easy solution is to calculate the log-likelihoods instead of the likelihoods (as we've done in figures above). Because the logarithm of a series of numbers multiplied together equals the sum of the logarithms of those numbers, the log-likelihood is

$$\log(L(r)) = \sum_n \log(P(x_n|r,t)) \tag{8.4}$$

For the data above, the negative log-likelihood is minimized at $r = 0.2$ geoducks m^2 (figure 8.4). I've also shown the range of r that have negative log-likelihoods that are less than or equal to the smallest negative log-likelihood plus 1.92. The 95% confidence interval based on these data is $(0.117, 0.315)$.

Learn how to perform these calculation in section 16.3 (pp. 275–80).

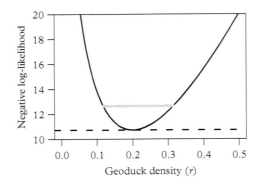

Figure 8.4 *Log-likelihood for different values of* r. *The horizontal line shows the maximum likelihood; the gray line shows the confidence interval.*

8.5 Advanced: Nuisance parameters

We often conduct a particular study or experiment to measure or estimate one parameter, but the probability density function has two or more unknown parameters that need to be estimated. For example, we may want to estimate the mean of a normally distributed random variable, but, to calculate likelihood, we also need to estimate the variance. Also, we may be primarily interested in one parameter of a model, say, the maximum population growth rate, but the model requires other parameters such as carrying capacity. We call these variables that need to be estimated but are not of primary interest "nuisance parameters."

There are basically two ways to deal with nuisance parameters. The first is to assume that we know the value of the nuisance parameter, that is, we set it to some value. This is not a great solution unless you have a lot of prior knowledge of what the variable ought to be.

The better solution, is to conduct what is known as a "likelihood profile." The basic idea is similar to what we have done so far, where we defined a range of values for the parameter of interest and calculated the likelihood for each value. The main difference is that there is now a second parameter. So, rather than calculating likelihood for a single list of parameters, we will instead define a list of the second nuisance parameter, and then calculate likelihood for each combination of the two parameters.

8.5.1 What is a likelihood profile?

A likelihood profile describes the likelihood (or log-likelihood, or negative log-likelihood) as a function of a parameter of interest, where the other parameters are adjusted to make the model fit as good as possible. Consider a model that requires a vector of parameters, called θ. We then partition that vector into the parameter that we're interested in—let's call that ψ—and all of the others, which we will call θ^*. We define the likelihood profile value as

$$L_{\text{profile}}(\psi)) = \max_{\theta^*} L(\theta^*, \psi; y) \tag{8.5}$$

What does this mean? First, the operation "max" with θ^* underneath it means "get the largest value that you can by adjusting θ^*, given the data, y, and a given value of ψ." So, this equation assigns a likelihood profile value to any value of ψ, as the largest possible likelihood you can get by adjusting all of the remaining parameters.

We know that we often like to work in negative log-likelihood space, so we can write this also as

$$-\log(L_{\text{profile}}(\psi))) = \min_{\theta^*} \left(-\log(L(\theta^*, \psi; y))\right) \tag{8.6}$$

This is likely confusing by now. Don't worry—it will hopefully clear up with an example.

Table 8.4 *Survey results of geoducks sampled over six different areas*

Area surveyed (m²)	Number of geoducks counted
20	2
20	1
40	7
40	1
40	2
20	5

8.5.2 Example

You conduct several surveys for geoducks (table 8.4). You are interested in estimating the mean geoduck density, but you are not comfortable assuming a Poisson distribution because geoducks are very patchy. Instead, you choose to use a negative binomial probability density function, which requires estimating the parameter k. We want a confidence interval in r, but, to do so, we must navigate the nuisance parameter k.

Learn how to perform these calculations in section 16.3 (pp. 275–80).

To begin, we specify a list of values of r that we will consider: 0.025 to 0.30, in increments of 0.0025.

For each, we want to calculate the profile negative log-likelihood, which is the smallest possible negative log-likelihood you can get by adjusting k, given each value of r. To make it more concrete, suppose we are calculating the profile negative log-likelihood for $r = 0.1$. We do that by finding the value of k that, when accompanying $r = 0.1$, makes the negative log-likelihood as low as possible.

Well, how do we do that? Here I'll show an approximate approach. This doesn't give the greatest accuracy in the world but is, frankly, the best way to *learn* the basics of what is going on, even if it isn't the best way to do it in practice.

In the approximate approach, you will pretend that k can only take discrete values. Here, we might consider k from 0.5 to 20 in increments of 0.5. This is why I refer to this approach as "approximate," because, truly, k can take any value.

Given this approximation, you cycle through the combinations of r and k, and calculate the negative log-likelihood for the observations, given each combination of r and k.

Figure 8.5 depicts a heat map showing the negative log-likelihood (purple means high, yellow means low) for each combination of r and k. The bright yellow swatch through

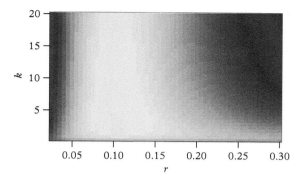

Figure 8.5 *Negative log-likelihood for each combination of r and k.*

the middle is the area that has the smallest negative log-likelihood. The negative log-likelihood is minimized when $r = 0.105$ and $k = 4$.

But we don't care that much about the parameter k—we mostly care about the parameter r. We need some way to collapse this two-dimensional likelihood profile into one dimension (the r dimension).

8.5.3   The likelihood profile

Remember that, for each value of r, we want to assign the smallest possible negative log-likelihood you can get. You take each candidate value of r above, look at the negative log-likelihoods calculated across the different values of k, and find the smallest negative log-likelihood. We'll call this the "profile negative log-likelihood," because it is the smallest negative log-likelihood found by profiling over our nuisance parameter. We repeat this for each value of r. We identify the maximum likelihood estimate and the confidence interval (figure 8.6) in the same way that we used negative log-likelihood for single parameter models. Here, the confidence interval is fairly wide (0.0525–0.2225), because it admits that the encounters could be generated by a patchy distribution and that we don't know the variance of the distribution.

Learn how to create more precise profiles in section 16.3 (pp. 275–80).

In summary, the likelihood profile is calculated as follows:

1. You specify a value for the parameter you are interested in.
2. You find the smallest negative log-likelihood that you can get, given this parameter value, by changing the other model parameters.

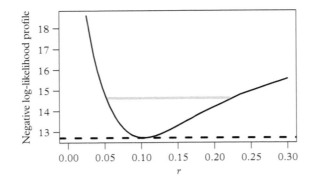

Figure 8.6 *Likelihood profile for geoduck density,* r, *profiling over values of* k. *The dotted line depicts the minimum negative log-likelihood; the thick gray line depicts the confidence interval.*

3. You assign this profile negative log-likelihood to the parameter value and repeat for many different values of the parameter of interest.

4. You then apply the same procedures as in section 8.2 to calculate a confidence interval.

8.6   Estimating parameters that do not appear in probability functions

So far, all of our examples estimated one (or more) parameters that appeared in a probability function. We estimated p in the binomial function. We estimated r in the Poisson and the negative binomial function. However, one often fits model parameters that do not appear in our probability density function. For example, you might want to estimate the parameters r and K of the logistic population model. How can we use likelihood to estimate these parameters?

All you need to do is mathematically relate the parameters of interest to other parameters that *do* appear in a probability density function. This is best illustrated through an example.

8.6.1   Entanglements of Hector's dolphins

Hector's dolphins are the smallest and rarest dolphins in the world. Their maximum size is only 1.4 m, so they might be smaller than you. They are endemic to New Zealand, and they are threatened by incidental entanglement in gill nets from coastal fisheries. Obviously, there is interest in revisiting fisheries regulations to reduce net entanglements. But, to do so, one needs a way to assess how much a change in fishing practices is needed to change entanglement rates.

We can model the relationship between the amount of fishing effort and the expected number of entangled dolphins through the following relationship:

$$\mathbb{E}[k] = qEN \tag{8.7}$$

where k is the number entangled over a period of time, E is the amount of fishing effort, N is the abundance, and q is called the catchability coefficient. It tells you the fraction of animals that are entangled with 1 unit of fishing effort. This is an important parameter to estimate. If the parameter of q is very high, that means that small changes in fishing effort will reduce entanglements. If the parameter of q is low, then larger changes are needed.

Martien et al. (1999) provide information on the number of Hector's dolphins killed, the amount of fishing effort, and the estimated population size, summarized in (table 8.5). You wish to estimate the catchability coefficient, q, so that you can model population dynamics under different fishing management scenarios.

First, we must identify our random variables. In this case, it is k, the number of dolphins killed in a given year. We recognize that this is a binomial random variable because it is a discrete outcome over a discrete number of trials. Our notation then is:

$$k \sim B(N, p) \tag{8.8}$$

where the notation $\sim$ denotes that the variable on the left is a random variable, drawn from the probability function on the right, in this case, a binomial probability mass function. We already know what N is; it is the number of dolphins initially present. But what is p, and how does it relate to q? We can do some quick rearranging of equation 8.7, using the expected value of the binomial distribution (section 7.2), to see that equation 8.7 predicts a relationship between q and the probability that an individual dolphin will be entangled:

$$\frac{\mathbb{E}[k]}{N} = p = qE \tag{8.9}$$

where p is the probability that an individual dolphin will become entangled; p is related to the parameter q by $p = qE$. Unlike q, the parameter p does appear in the probability mass function.

Table 8.5 *Summary of Hector's dolphin catch data*

	1985	1986
Dolphins entangled (k)	79	70
Effort (1,000 m^2 km^{-2} d^{-1})	0.267	0.213
Population size	1024	990

This is good news, because it means that we can apply the same procedures as we used before to estimate q with likelihood, with only a simple intermediate step that relates our parameter q to the parameter p. So, the steps are as follows:

1. List candidate values of q, ranging from 0.2 to 0.45, in small increments.
2. For each q, calculate the associated p for each year, by multiplying q times the effort in each year. This gives us a separate value of p for each year.
3. Calculate the negative log-likelihood of each year's observation, using the p for that year.
4. Sum the negative log-likelihoods of each observation to get the overall negative log-likelihood.
5. Identify the maximum likelihood value of q, and use that to generate a confidence interval.

Applying these steps, you should get a maximum likelihood estimate of q equal to 0.307, with a confidence interval of 0.262–0.357 (figure 8.7).

See these calculations in more detail for Excel and R in section 16.1 (pp. 263–273).

Basically all parameter estimation problems will follow analogous steps. You link your parameters to those that appear in the appropriate probability function and then apply the same steps we used earlier to find the maximum likelihood estimates and to calculate uncertainty via confidence intervals.

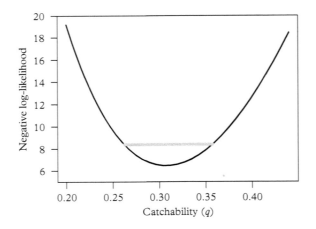

Figure 8.7 *We can calculate log-likelihood for parameters that do not appear in the probability density function, like the catchability coefficient.*

8.6.2 Practice example 2

Often we wish to fit specific functions of our dynamic models to data. For instance, our various population growth models (section 2.4.3) all contain functions that we might want to estimate from data. We can use maximum likelihood estimation to estimate parameters of those models, as long as we can relate those models to one or more parameters of a probability density function.

As an example, Forrester (1995) conducted an experiment by placing a small goby (*Coryphopterus glaucofraenum*) in coral reefs at different densities. Survivorship in these reefs is likely density dependent because there are limited refuges from predation, or because high densities expose the fish to parasites. Because these fish do not move very much, the author could revisit these sites after about ten weeks to see how many remained.

The data consist of several reefs of roughly equal area, each stocked with a different number of individuals. The data are illustrated in table 8.6. You wish to fit the following model:

$$\mathbb{E}[N_{\text{survive}}] = N_0 p(N_0) \tag{8.10}$$
$$p(N_0) = p_{\text{max}} e^{-\alpha N_0}$$

where N_0 is initial density, $p(N_0)$ is the survivorship function that depends on N_0, p_{max} is the maximum survival probability, and α describes the rate at which survivorship declines with N_0. Note that this is a variant of the Ricker model that we explored earlier in the book (section 2.4.3). We wish to estimate the parameters p_{max} and α from these data. As before, we need to identify our random variable and the associated probability function: here the random variable is N_{survive}, which is drawn from a binomial distribution: $N_{\text{survive}} \sim B(N_0, p)$. The steps are as follows:

1. Choose the appropriate likelihood function. This is a binomial probability mass function: we are modeling the number of discrete individuals that survive.

2. Relate parameters to those that appear in the probability function. Here, I've set up the equation to illustrate this link clearly, that is, the function $p(N_0)$ gives the binomial p for each value of N_0. For each experimental reef, calculate p, given the initial number of gobies, N_0, for any choice of p_{max} and α.

3. Calculate the negative log-likelihood for each observation, using a single combination of candidate values of p_{max} and α. Sum the negative log-likelihoods over all observations to get the total log-likelihoods.

4. Choose your favorite method to find the maximum likelihood estimates (you can use one of the numerical methods described in the skills section, or loop through combinations of model parameters).

See this calculation in section 16.2 (pp. 273–275).

Table 8.6 *Goby survival results*

Initial number of gobies	Final number of gobies
6	4
10	4
12	4
20	7
29	9
43	8
59	18
91	11
22	9
25	8
8	3
12	7
15	5
42	17
68	16

Source: Data from Forrester (1995).

Figure 8.8 shows the fit of the model at the maximum likelihood parameter values of $p_{max} = 0.5$ and $\alpha = 0.0127$.

The main point is that once you've properly identified your random variables and specified the appropriate probability function, all that is left is to link the model parameter you are estimating to the parameters of the probability function. This means that you can fit any parameter of any model you want, provided you have the right data on hand.

8.7   Estimating parameters of dynamic models

Part 1 of this book focused on the development and application of dynamic models. Part 2 has explored parameter estimation (and will explore model selection later), but all of those examples were static: there was no time dynamic involved in those models. Here I briefly review issues and concepts associated with fitting dynamic models to time series data.

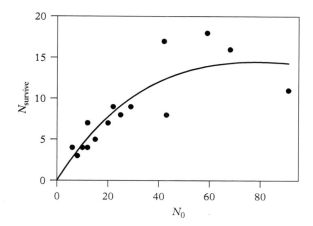

Figure 8.8 *Initial and surviving number of gobies on reefs. Line shows the model at maximum likelihood parameter estimates. Data from Forrester (1995).*

Presume we have a series of observations of population abundance $N_0, N_1, N_2, \dots, N_t$. We want to use these data to estimate the parameters r and K of a logistic population model. We therefore presume that the logistic equation gives the expected number of individuals seen in time N_t, and that the data are generated from a some probability function. One way to model the data is to assume that the observations are perfect but that there is randomness (stochasticity) in the population dynamics process itself:

$$\mathbb{E}[N_{t+1}] = N_t + rN_t\left(1 - N_t/K\right) \tag{8.11}$$

where N_{t+1} are random variables drawn from some appropriate probability function, and, importantly, N_t are the observations that are presumed to be measured without any statistical error. For this reason, we call this a process error model, because randomness occurs in the model process itself.

We might assume that the random variables are normally distributed:

$$N_{t+1} \sim N(\mu, \sigma^2) \tag{8.12}$$

and because we know that the expected value of a normal distribution equals μ, we can therefore link the parameters of our model, r and K, to the parameters of the probability function, using equation 8.11. In this case, we would have three parameters to estimate: the population model parameters, r and K, plus the standard deviation of the process error, σ.

One could use many different probability functions for the random variables. Lognormal is sometimes used because the standard deviation scales with the mean, which is common in ecological data. If the observed population sizes are discrete counts, you might prefer a discrete probability mass function like the Poisson or the negative binomial. For instance, if you assumed a Poisson,

$$N_{t+1} \sim \text{Pois}(\lambda) \tag{8.13}$$

and because we know that the expected value of the Poisson distribution equals λ, we use equation 8.11 to generate λ for each observation.

An alternative way to fit the model is to assume that the population growth is deterministic (i.e., the population dynamics are perfectly described by the population model), and the observations are subject to random error. If you made these choices, then the model would be

$$N_{t+1} = N_t + N_t r(1 + N_t/K) \tag{8.14}$$
$$\mathbb{E}[N_{t,\text{obs}}] = N_t$$
$$N_{t,\text{obs}} \sim N(\mu, \sigma^2)$$

where, once again, we are assuming a normal random variable, but here the randomness is in the observations, not the population dynamic process. That is, $N_{t,\text{obs}}$ are the random variables. For this reason, we call this an observation error model. As before, we use the fact that the expected value of a normal distribution equals μ, to relate the population model parameters to the parameters that appear in the probability function.

In each case, you would apply maximum likelihood methods to solve for r, K, and σ, but in very different ways. In the process error model, you would use each observed N_t to predict each N_{t+1}, and then calculate the likelihood of the observed N_{t+1} based on this prediction. In other words, the process error model only projects one year ahead for each pair of N_t and N_{t+1}. In the observation error model, you would project out the predicted abundance for all years $0, 1, 2, \ldots, t$, via the logistic model (using assuming some estimated starting abundance, $\widehat{N_0}$), basically in the same way as is shown in section 12.1. You would then calculate the likelihood based on these predicted values.

Learn how to estimate parameters using process or observation error in chapter 17 (pp. 283–94).

So, which should you use? Neither is perfect, and the performance of either can be quite awful (Polacheck et al. 1993). My guidance is to at least respect the data: when the data consist of time series of population abundances and you are fitting a model that projects population dynamics, at the very least examine the time dynamics to make sure you are fitting a reasonable model.

For illustration, consider the long-term time series data of house wren counts from a preserve in Illinois (figure 8.9), which is based on data from Kendeigh (1982). The panel

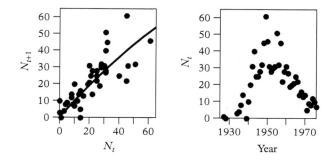

Figure 8.9 *Two views of the same house wren time series. The fitted process error model (left) can't account for the true time series dynamics (right).*

on the left shows the time series as pairs of N_t and N_{t+1}, with the best-fitting process error model shown as a line. That analysis suggests evidence for density dependence, as the model fit indicates a nonlinear (curved) relationship between N_t and N_{t+1} (Pollard et al. 1987). The panel on the right shows the same data as a time series. The data show a period of population increase for the first twenty years, a period of leveling off, and then a period of steep decline. The logistic model cannot predict this behavior! Clearly, something else that is not captured in the logistic model is driving this population. This should make us deeply suspicious of the model fit in the left-hand panel. How do we know that the apparent leveling off of population abundance in the mid-1940s was not just a harbinger of the deteriorating external forces that eventually caused the population to decline?

You might rightly ask, "Why don't we include both process and observation errors?" True, we think both are present: we don't believe that our observations are perfect or that our model perfectly predicts the abundance from one time step to the next. For quite a while, estimating both process and observation error was intractable. Fortunately, the development of "state-space" models allows one to estimate both simultaneously (Schnute 1994), and there has been quite a bit of advances to make these tractable for everyday ecologists (Valpine and Hastings 2002). Still, these are fairly advanced models and are well outside the scope of this book.

8.8 Final comments on maximum likelihood estimation

In all of the examples so far, we estimated maximum likelihood parameter estimates and confidence intervals by assuming a list (one parameter) or an array (two parameters) of alternative parameter values. This simplification was a useful learning method, because it makes clear the underlying concepts and applications.

There are two obvious limitations of this approach. The first is that we often have more than two parameters to estimate. In fact, it is hard to find examples where there are two or fewer parameters to estimate because even for models that have two parameters, we often have additional nuisance parameters (like the variance) to estimate. The second is that it forces us to define a priori what parameter values are candidates for the maximum likelihood. By listing the parameters and calculating likelihood for only those parameters, we may be missing the truly best possible parameter.

For these reasons, we usually use numerical methods to search for the combination of parameter values that maximize the likelihood. In Excel, you can use the add-in Solver, which is actually quite good. In R, you have a lot of choices.

Learn how to apply numerical optimization methods in section section 16.1 (pp. 263–73).

8.9   Overdispersion and what to do about it

Sometimes the probability functions that we've selected cannot adequately account for all of the variation in the data. This is particularly problematic when the variance of the sampling distribution is determined by the same parameters that govern the mean. For instance, in the binomial and the Poisson, the variance is "set" by the same parameters that explain the mean.

Overdispersion can be a major issue in ecological data. Rarely is it true that binomial trial probabilities are identical over all trials or that the density of some organism is constant over some sampling frame. In these cases, you run the risk of greatly overestimating the precision of your model parameters if you assume a model that has less variance than the true process that generated the data. Also, overdispersion can greatly impact the model selection procedures described in the next chapter (where it imposes a bias towards selecting more complex models).

How can you reliably detect it? To start, try some reasonable things like comparing the sample variance to the estimated variance from the fitted probability density function. Make plots of the frequency distribution of the data, and compare those to the best-fitting probability distribution. If these are very different, you probably have overdispersed data and need to select a different probability mass function (the beta-binomial for counts over discrete trials, or the negative binomial for counts over continuous sampling). You may also need to try one of the zero-inflated models (section 7.7).

There are, of course, more rigorous ways of addressing and assessing overdispersion, but these require concepts like deviance, and sampling distribution of deviance ratios, that are beyond the scope of this book. Nevertheless, it is an extremely important topic once you start fitting models to ecological data, so be sure to consult more advanced texts such as those by Bolker et al. (2008) and Burnham and Anderson (1998).

Summary

- If we have some observed outcome (data), we can estimate parameters of the probability density function.
- The likelihood of a parameter value is the probability of the data, given that parameter value.
- The "maximum likelihood estimate" is the estimate under which the data are most likely to occur.
- We can estimate parameters, ecological models as long as we can link the model parameters to those of some probability density function.
- Likelihood ratios can be used to estimate the precision of parameter estimates.

Exercises

1. Work through the example in section 8.3.3 to estimate geoduck density.
2. Work through the example in section 8.6.2 to estimate p_{max} and α.
3. Below (table 8.7) are data from Green and Peloquin (2008), who conducted an experiment to test how exposure to low pH affects survival of larval stages of the southern two-lined salamander (*Eurycea cirrigera*). Each experiment used ninety-six salamanders. You wish to estimate the pH level at which 50% of the larvae die. To do so, you fit a model whereby the probability of dying equals $p = \frac{1}{1+\exp(-(1+(x-x_{50})\beta))}$, where x is the pH, x_{50} is the pH level at which 50% mortality occurs, and β is the slope describing the rate at which mortality declines with pH. Find the maximum likelihood estimate of x_{50}.

Table 8.7 *Experimental results*

pH	Number of deaths
2.75	96
3.75	84
3.25	6
4.25	2
4.75	1
6.5	1

Source: Data from Green and Peloquin (2008).

4. Advanced: Calculate the 95% confidence interval for x_{50} from exercise 3 above.

9

Model Selection

We motivate this chapter with a brief example: the case of the Western monarch butterfly (*Danaus plexippus plexippus*), which, over the past two decades, has experienced a 97% decline from its historical average abundance and declined 86% from 2017 to 2018 alone (Pelton et al. 2019). This is obviously a critical conservation concern, which requires solutions that address root causes. Undoubtedly, there is more than one cause—indeed, overwintering habitat loss and pesticide use are both believed to be important contributors to the decline (Pelton et al. 2019). Given the multiple root causes of this and many conservation challenges, we wish to adopt a hypothesis-evaluation framework that allow us not only to consider multiple alternative hypotheses simultaneously but also to measure the relative weight of evidence in support for each. That is, we seek to measure *degrees of support* for alternative hypotheses and thereby admit the possibility that one or more hypotheses might have equivalent support. When that happens, such as in the case of the the Western monarch butterfly, it tells us either that multiple policy options are needed to address distinct root causes or that the data are insufficient to distinguish among some competing hypotheses.

Put another way, when we talk about model selection, we are really talking about weighing evidence for alternative hypotheses expressed as mathematical and statistical models. This means we need a way to measure "support" for a hypothesis. The methods presented here rest on two key principles:

1. A model is well supported if it makes good predictions. That is, we judge the weight of evidence for a model by how well it predicts future observations.

2. We judge prediction ability by how likely the future data are, given the model. If the data are highly likely under the model, then the model has good predictive ability. If the data are highly unlikely under the model, then the model has poor predictive ability.

In other words, we have a bundle of data. We will fit model parameters to those data to create a fitted model and then repeat that process for each alternative model. We will judge the models by the quality of their predictions, that is, by calculating a metric that judges how well each model predicts things (Konishi and Kitagawa 2008).

Introduction to Quantitative Ecology: Mathematical and Statistical Modelling for Beginners. Timothy E. Essington, Oxford University Press. © Timothy E. Essington 2021.
DOI: 10.1093/oso/9780192843470.003.0009

Model selection using null hypothesis testing

Null hypothesis testing can also be used to choose between two models, but uses an entirely different framework to do so. Consider two models, $f(x)$ and $g(x)$, both fit to the same data x, where one model is a simpler version than the other. Suppose $f(x)$ explains the Western monarch butterfly decline based only on overwintering habitat loss, while $g(x)$ explains the decline based on both habitat loss and pesticides. As for all null hypothesis tests, we either accept or reject the null hypothesis if a test statistic is unlikely if the null hypothesis is true. In this case, the null hypothesis is that $f(x)$ is true. The test statistic is the log-likelihood ratio between the fitted models $f(x)$ and $g(x)$ (log-likelihood ratio literally means the log of the ratio of likelihoods). If the null hypothesis is true, this ratio follows a chi-square distribution with degrees of freedom equal to the number of additional parameters in the complex model. In mathematical terms,

$$R = \log\left(\frac{L(f)}{L(g)}\right) \tag{9.1}$$

$$2R \sim \chi^2_{df=k_g-k_f} \tag{9.2}$$

where k_i is the number of parameters in model i. Thus, the steps are to calculate the maximum likelihoods for each model, calculate the log-likelihood ratio, and then ask, "What is the probability of getting a ratio equal to or greater than what I observed, if the null hypothesis, $f(x)$, were true." If this probability is small, we reject the null hypothesis. If it is large, we accept the null hypothesis that the simple model is "right."

9.1 Framework

We presume that there is some true distribution or model that exists that generated the data that we see. We obtain a sample of random variables $x_1, x_2, x_3, ..., x_n$ generated by that true process. We then fit a model, denoted $f(x|\theta)$ with parameters θ to these data, to find the value of the parameters that maximize the log-likelihood of the data. The fitted model is one in which we replace the unknown parameters θ with the maximum likelihood estimates, denoted $\hat{\theta}$.

We want to know whether this fitted model, $f(x|\hat{\theta})$, provides good predictions for future data. Presume that we repeated the experiment or sampling procedure and collected a new set of data in the same way that we collected the original data. On average, what would be the log-likelihood of our fitted model to these new data?

Let's make this more concrete by considering the Hector's dolphin entanglement data described in section 8.6. Your data are observations of the number of entangled dolphins in each of the two years. Your model, $f(x|\theta)$ is the binomial probability function where the binomial probability p is given by q times E, x is the number of dolphins entangled, and N is the number of dolphins in the population in each year. You find the value of q, $\hat{q}$ that maximizes the log-likelihood.

Now suppose data continue to be collected in 1987 and 1988, and, by miraculous chance, the population sizes and fishing efforts were identical to those seen in 1985 and 1986. How well will this model predict these future observations? In other words, what is the expected log-likelihood when applying the fitted model to the new observations? If this value is very high, then this means that future observations are very likely under the model and that therefore the model has predicted those observations rather well. Thus, we use the expected log-likelihood to judge prediction ability.

So how do we calculate the "expected log-likelihood"? On one hand, it might be tempting to simply use the log-likelihood at the maximum likelihood estimates. In other words, if you want to know the expected log-likelihood for future observations, you might consider using the log-likelihood from the observations you have in hand. When you fit q to the Hector's dolphin data, you will find that the maximum likelihood estimate, $\hat{q}$, is 0.307, with a log-likelihood of -6.45. So perhaps we can presume that the expected log-likelihood of two future observations given $\hat{q}$ equals -6.45.

It turns out that the maximum log-likelihood of any fitted model is biased: it tends to be larger than the expected log-likelihood applied to new observations. This is because the model-fitting procedure will naturally bend the model to fit the particular set of data (random variables) that you have. In the goby example (section 8.6.2), perhaps there was one experiment that by chance happened to have a particularly low survivorship. The model will to try to accommodate that observation when calculating the maximum log-likelihood, bending the model predictions to more closely match the observation. This tendency to overestimate the expected log-likelihood becomes greater as the model has more parameters, because these additional parameters allow the model to bend even further to chase the data.

So calculating the "expected log-likelihood" seems like a good basis for model selection, but it seems challenging to know what this value is.

Bias-precision trade-off

If you explore model selection and model complexity, you will undoubtedly come across the idea that there is a trade-off between bias and precision. A good model is one that is both unbiased (meaning that, on average, it gives the correct prediction) and precise (meaning that, if you repeated the estimation procedure with new data, you would likely get the same predictions). Yet, measures taken to reduce bias also decrease the precision.

You can see this in figure 9.1. Here I generated a random data set of twenty-five observations of some independent variable, X, and some dependent variable, Y. The simplest model I can fit is

$$Y \sim N(\mu, \sigma^2)$$

continued

which is only fitting the mean and variance of the Y, ignoring the independent variable X entirely. This produces very biased predictions (see figure 9.1, upper left panel), consistently over-predicting Y when X is small and underpredicting Y when X is large. However, when I simulate the collection of these data and fits several times, I generally get similar (bad) predictions each time (figure 9.1, lower left panel). Alternatively, I can fit an overly complex model by making μ a sixth-order polynomial in X:

$$\mu = \sum_{i=0}^{6} \beta_i X^i$$

This produces unbiased predictions (figure 9.1, upper right panel) as the fitted μ closely match the data. But when I repeat this many times, the predicted μ varies enormously (particularly at the low and high values of X; see figure 9.1).

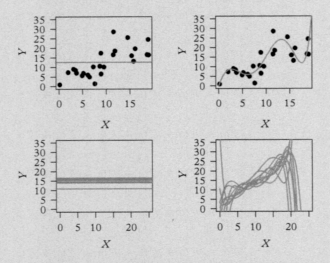

Figure 9.1 *High bias is caused by overly simple but precise models (left panels), while poor precision is caused by overly complex but unbiased models (right panels).*

For this reason, a lot of the descriptions about model selection refer to optimizing this trade-off: striking the balance of bias and precision.

9.2   An intuitive method: Cross validation

Suppose you have a lot of data on hand. So much that you can reliably fit the parameters of each model based on only half of all of the available data. In that case, you might do just that, using one-half of the data as the so-called training data to fit the parameters of each alternative model. Then, you can apply the fitted models to the remaining data, usually called the "validation data," and measure the log-likelihood of the validation data given each fitted model. If your sample size is big enough, this sample-derived quantity should be relatively close to the expected log-likelihood.

Cross validation has obvious appeal. It is simple and intuitive, and it directly measures the very thing that we wish to use to judge model support. So, why don't we use cross validation all the time? Mostly because we often don't have sufficient data to do it. For the method to produce reliable estimates of expected log-likelihood, you need fairly large sample sizes in both your training and your validation data set. In many (most?) cases, we do not have that luxury, so we must instead use other methods.

9.3   The Akaike information criterion as a measure of model performance

Information criteria provide an alternative to cross validation: they use information about the model fit and model complexity (number of parameters) to estimate what would have happened if you could perform a cross validation (Anderson et al. 2000; Burnham and Anderson 1998). The very first information theory was derived by Hirotugu Akaike (Akaike 1974), who discovered that, under certain conditions, an unbiased estimator of the expected log-likelihood equals

expected log-likelihood = maximum log-likelihood

$-$ number of estimated parameters in the model

To "take historical reasons into account," he multiplied this difference by negative two to generate the "Akaike information criterion" (AIC):

$$AIC = 2(-L_{max} + K) \tag{9.3}$$

where $-L_{max}$ is the negative log-likelihood at the best-fitting (maximum likelihood) parameter values, and K is the number of *estimated* parameters. The AIC tells us the *relative* predictive ability of our alternative models, where models with the lowest AIC will give the better predictions. Thus, this very simple expression, which uses calculations that you already have on hand when fitting model parameters to data, provides a quantitative measure of the degree of support that the data give to a particular model.

We should take a moment and look at its components. We know that a good model should have a low score (because of the multiplication by -2). A model that fits the data well will tend to have a high likelihood, which in turns generates a small negative log-likelihood. AIC also penalizes models for each parameter that is estimated. The best-fitting model therefore has a small negative log-likelihood with the smallest possible number of parameters.

Because AIC only gives the relative predictive ability of a model, the actual AIC value for any individual model is useless. You can't say that AIC = 10 is good. Rather, we compare AIC across the alternative models. If we have five candidate models, each of which is fit to a given set of data, first we calculate AIC and then the differences in AIC between the models. Models with small AIC are better than those with large AIC, so we often compare all candidate models to the model with the smallest AIC:

$$\Delta AIC_i = AIC_i - \min_i AIC_i \qquad (9.4)$$

where i denotes a model.

Fun fact

Today we say that AIC stands for "Akaike information criterion," but Hirotugu Akaike himself never intended for it to be named after him. He coined the term AIC, thinking that IC stood for information criterion, and he added the prefix A to denote that this was the first defined information criteria "so that similar statistics, BIC, DIC- etc., may follow" (Akaike 1974, 719).

The simplicity of the AIC calculation allows you to see how it directly addresses the bias-precision trade-off. That is, models that explain the data well have a lower AIC because they have smaller negative log-likelihood, but this is balanced by the number of parameters that were fit to achieve that negative log-likelihood. Burnham and Anderson (1998) emphasize that this simple outcome belies a fairly sophisticated application of mathematical statistics; it is not an arbitrarily defined way to balance the bias-precision trade-ff.

9.3.1 Take-home points

- Model selection often works by evaluating several models and determining which provide the best predictions.
- We judge predictive ability based on on the "expected log-likelihood" of the fitted model to new data.
- AIC is a simple way of calculating "expected log-likelihood."
- ΔAIC gives us a measure of the relative performance of candidate models.

9.3.2 Advanced: Theoretical underpinnings of information theory

We know that all models are simplifications of reality. Presume that there is some "truth" that generates our data. Truth is the full set of environmental, ecological, behavioral, and statistical processes that underly the random variables that comprise our data. We want to represent these data with different models, where each model corresponds to a different hypothesis about how the real world works. Each model is different from truth (Aho et al. 2014) and can be thought of a having some distance away from truth (figure 9.2). In information-theory circles, this is called the Kullback and Leibler distance, or K-L distance for short (Kullback and Leibler 1951).

9.3.2.1 *What is the distance from truth?*

The distance between the model and truth is the "information loss" that comes from representing truth with a simplified model. Information loss is essentially how inaccurately a simplified model predicts the probability of observations. A good model is one that minimizes this information loss, while a poor model has a high information loss. In the schematic above (figure 9.2), model 2 looks to be closest to the truth, while model 4 is the furthest.

But what exactly is "information loss," according to the K-L distance? Consider a model that is predicting the outcome of a random variable, x. That prediction is usually a probability distribution, which we denote $g(x)$. The true probability distribution, that is, the truth, is another function, which we denote $f(x)$. The K-L distance measure is

$$D(g|f) = \begin{cases} \sum f(x) \log\left(\frac{f(x)}{g(x)}\right) \text{ if } x \text{ is discrete} \\ \int f(x) \log\left(\frac{f(x)}{g(x)}\right) dx \text{ if } x \text{ is continuous} \end{cases} \tag{9.5}$$

There is a catch, however. *We have no idea where truth is.* That is, we don't know the true generating distribution $f(x)$. How can we measure the distance between our models

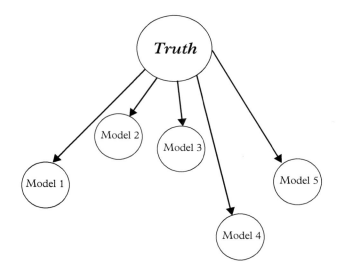

Figure 9.2 *Information theory seeks to find which model is closest to some unknowable truth.*

and truth when we have no idea how much information we are losing with each model? Well, we can't. But we don't need to know exactly how much information we are losing, only the *relative* amount of information we are losing. That is why the AIC by itself is meaningless. It is a relative measure of information loss. It only becomes useful when compared to other AIC scores.

A useful analogy

Consider a marathon race. We want to know which of the runners is most similar to the fastest runner there ever has been and ever will be. We don't know who the fastest runner is, or how fast they run. But it is safe to assume that whoever finished the marathon first is probably closer to *the fastest ever* than the person who finished second and is certainly closer than the person who finished last. We know this because we have a metric, the time it took to complete the marathon, that we can compare across the runners who competed in the marathon. In fact, it is common to report an individual runner's marathon times as the time behind the winner. If we see that someone was only five seconds behind the winner, we know that they are nearly as close to the *fastest ever* as the winner of the race. We can confidently say that the winner is much closer to the *fastest ever* than someone who finished one hour after the winner.

   We use a similar reasoning in information theory. We want to know the best model, one that is closest to the truth. We can generate a quantity, similar to the marathon time, that allows us to rank the models, even though we don't know what the truth is. We do this much in the same way that we can rank the marathon runners even though we don't know what the *fastest ever* runner could do.

9.3.3   Alternatives to the AIC

While using the AIC has certainly taken hold as a common way to perform model selection, it is worth noting that there are other ways of getting at the question of "which is the best model for future predictions."

   In fact, it turns out there are many different information criteria. A common one is BIC (also sometimes called the Schwarz information criteriaon [SIC]), which looks similar to the AIC but calculates the expected negative log-likelihood in a different way:

$$BIC = -2\log L_{\max} + K \log n \tag{9.6}$$

where the first term is again the negative log-likelihood at the maximum likelihood parameter estimates, and n is the sample size. Aho et al. (2014) give an excellent summary of the different assumptions and uses of the BIC over the AIC. But keep in mind that there are many other information criteria—see the work by Hooten and Hobbs (2015) and Burnham and Anderson (1998) for more examples.

9.4 Interpreting AIC values

We know that the model with ΔAIC equal to 0 is the best model. We know that models that have really large ΔAIC give worse predictions and are therefore less supported. Ultimately, we want to say whether certain models have sufficient support to warrant any further consideration. In other words, how big does a model's ΔAIC have to be before we dismiss it as being unlikely?

It turns out that there are no firm rules. Unlike null hypothesis testing, where, by convention, we use $\alpha = 0.05$ as our threshold for determining statistical significance, judgments about model selection using the AIC are not as clear-cut. This is partly a good thing, because who is to say that a ΔAIC of 2 means that there is support but a ΔAIC of 2.1 means there is no support. In other words, we aren't going to apply hard threshold values. Rather, we acknowledge that degrees of support form a continuum, and our inference and interpretation should be consistent with that.

Still, by convention, the field has developed rules of thumb for interpreting ΔAIC scores. A good practice is to apply multiple rounds of screening. The first examines the model set and only looks at ΔAIC. The rules of thumb for this first round of screening are as follows (Burnham and Anderson 1998):

- ΔAIC $\leq$ 2: have substantial support and should not be excluded from further consideration
- ΔAIC between >4 and 7: have little support and might be excluded from further consideration
- ΔAIC > 10 have essentially no support and certainly should be excluded from further consideration

One possible outcome of the first round is that one model is far and away superior to the others (e.g., all ΔAIC but that of the best model exceed 10). In that case, you are nearly done (after examining goodness of fit; see below [section 9.4.2]) as the data give overwhelming support to one of your models.

In many other cases, though, you might have multiple models that have ΔAIC in the "plausible set." This means your work is not done! You now need to apply a second round of screening to decide which of those models with the "plausible set" should remain.

Here we will discuss two key issues for this second round of screening: support for simple models that are nested within complex models, and goodness of fit.

9.4.1 Nested versus nonnested

Suppose we wanted to use the killer whale time series data to judge evidence for three different population models: the density-independent model (equation 2.8), the logistic density-dependent model (equation 2.14); and the Beverton-Holt model (equation 2.19). The data, plotted as N_{t+1} versus N_t, are shown in (figure 9.3).

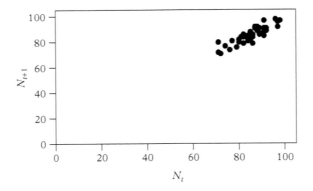

Figure 9.3 *The killer whale time series data plotted as* N_{t+1} *versus* N_t.

Table 9.1 *AIC table for population model comparison*

Model	Negative log-likelihood	Number parameters	AIC	ΔAIC
Density independent	140.46	1	282.91	0.00
Logistic	140.44	2	284.87	1.96
Beverton-Holt	140.454	2	284.91	2.00

We fit all three models, calculate AIC for each (table 9.1), and find that all models are within two units of each other (note: when showing AIC results for model selection, typically only the number of parameters and ΔAIC are presented, because they provide the information for inference, but here I included negative log-likelihood and AIC so that the calculation steps are better illustrated). This means that all survive the first round of scrutiny. The second round of scrutiny asks, "Are some of these models more complex versions of other models in the set?" In this case, the answer is yes. The logistic model becomes the density-independent model when K becomes very large (but remember K comes from assuming a linear density-dependent decline in per capita production with a slope β, so this implies the density-dependent slope equals 0). Also, the Beverton-Holt model becomes the density-independent model when β equals 0. This means the density-independent model is nested with the logistic and the Beverton-Holt models. The logistic and the Beverton-Holt models, however, are not nested within each other; there is no way to turn the logistic model into the Beverton-Holt model.

Notice that adding parameters to the density-independent model did nearly nothing to the negative log-likelihood (table 9.1). That means the data are equally likely to have occurred under all three models. Given this observation, it seems silly to keep the logistic model and the Beverton-Holt model in the set of supported models. Even though their ΔAIC was less than 2, it is clear that the data do not support the additional complexity of these models, given the performance of the simpler density-independent model. In this case, we would remove the logistic model and the Beverton-Holt model from the set of models that are supported by the data.

This was a pretty clear-cut case. But, more generally, how do you decide what level of ΔAIC provides sufficient support for the more complex model? We would probably only consider the more complex model if it had an AIC at least equal to that of the simpler model. For instance, if we applied the logistic model, and its AIC exactly equaled that of the density-independent model, we would conclude that the data support both models equivalently, that is, the data do not provide sufficient information to allow us to definitely discriminate between the density-independent model and the density-dependent model.

9.4.2   Fit to data

Just because models are supported on the basis of the AIC doesn't necessarily mean they are doing a good job at explaining the data or provide a good basis for predicting future observations. That is because we want to evaluate whether the models fit the data well. A good model makes predictions about observations that generally follow the observed patterns in the data. Poor fits can happen when the data strongly suggest a shape, trend, or relationship that the model cannot produce.

Here is an example. Arcese and Smith (1988) tracked clutch size, nest success, and the number of song sparrow offspring that survived to become fully independent individuals. They did this across a range of female densities to measure density dependence in reproduction. We can fit a density-independent model to these data (essentially, a straight line with intercept equal to 0) or a Ricker function, apply the AIC, and then ask whether the data support the complex model. figure 9.4 shows the data and fit of the Ricker model, and table 9.2 provides the AIC table. The ΔAIC values give overwhelming support

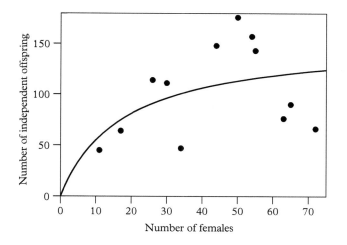

Figure 9.4 *Arcese and Smith song sparrow data, with best fitting Ricker model. Data from Arcese and Smith (1988).*

Table 9.2 *AIC table for song sparrow model comparison*

Model	Negative log-likelihood	Number of parameters	AIC	δAIC
Density independent	188.8	1	379.6	156.0
Ricker	109.8	2	223.6	0.0

Table 9.3 *AIC table for song sparrow model comparison with negative binomial distributions*

Model	Negative log-likelihood	Number of parameters	AIC	δAIC
Density independent/negative binomial	63.3	2	130.6	6.6
Ricker/negative binomial	59.0	3	124	0

for the density-dependent Ricker model. Does this mean that the Ricker model is well supported by the data?

Not really. The best fitting model has a very hard time capturing the actual trends in the data (figure 9.4). You can see that it generally overestimates offspring at low densities (points are below the predicted line), it underestimates at moderate densities (data points are above the line), and it overestimates again at very high densities. While the Ricker is better than a straight line, it is still not a great model for these data (imagine seeing the points without the curve—do you think that the fitted line is a good way to represent those data points?). We need a different function that can better capture the relationship between sparrow density and reproductive success, or at least capture the possibility that reproductive success drops off dramatically at high sparrow densities.

Even worse, the data are likely overdispersed. Consider some of the observations at high female densities. We can calculate the probability of these outcomes under the best-fitting model. If a model is fitting well, we expect the observations to be somewhat ordinary. For sake of illustration, I took the four data points associated with the four largest female densities (figure 9.5). For nearly all of these, the observed outcome was very unlikely! In some cases they are associated with a probability mass very close to 0. *The AIC is biased towards more complex models when the data are overdispersed.*

Indeed, such an observation would likely spur us to consider a negative binomial probability distribution. Not surprisingly, a negative binomial distribution fit the data far better (table 9.3), as judged by the much smaller negative log-likelihood, with only one additional parameter, compared to the Poisson models. Moreover, the difference between the linear model and the Ricker model became much smaller when we better accounted for the overdispersion in the data (table 9.3). This illustrates the bias in model selection when models are overdispersed.

The important point is that the AIC is not intended to be applied thoughtlessly without attention being paid to model fit. It would be misleading to report the AIC table and claim

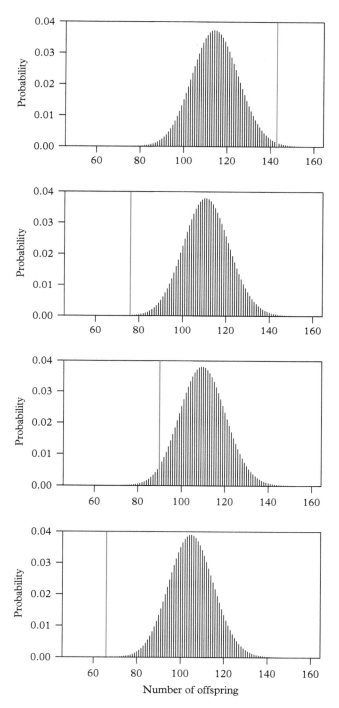

Figure 9.5 *Fitted probability mass distributions for the rightmost four data points in figure 9.4. Red lines show the observed outcomes.*

the Ricker model was best supported by the data, without also showing the troubling fit of this model to the data.

9.5   Final thoughts

9.5.1   Model selection practices to avoid

Model selection practices to avoid include the following:

- identifying the model with the lowest ΔAIC, claiming that it is the best model, and then dismissing all others even if they have relatively small ΔAIC: one main reason to apply information criteria is to find all of the models that have equivalent support
- blindly using a ΔAIC cutoff of 2 and applying a binary decision to each model, that is, supported versus not supported: the AIC is a continuous measure of degree of support
- forgetting that models are sometimes nested: when a simpler model is nested within a more complex model, the standard for evidence of the complex model is higher (usually, the AIC of the complex model needs to be greater than or equal to that of the simpler model)
- comparing models that are fit to different data sets: remember that the AIC is a estimation of what the resulting log-likelihood would be if you were to repeat the sampling in the same way so, an AIC fit to one set of data is not comparable to an AIC fit to a different set of data
- fitting every possible model that you can think of (see "True Story" below).
- ignoring model fit and overdispersion: just because a model has the most support does not mean that it is a particularly good model; do you think the model provides decent predictive ability?
- reporting only the AIC for a single model, and not the ΔAIC for all of the models
- combining information criteria and null hypothesis selection in the same analysis: remember that these are two completely different ways of drawing inference

9.5.2   True story

A team of scientists tested whether conservation areas are effective by asking whether a fish species is more likely to be present in a conservation area than in a paired area that is not protected. To predict fish presence, they used six variables (including degree of protection, plus several habitat related variables). They then tested alternative models that included from one to all variables, all possible interactions between variables, and second-order (interactive) responses to certain variables. The result was an analysis comparing 493 models. They used ΔAIC to find the best model or models.

9.5.2.1   *What is wrong with this scenario?*

At first glance, it might seem reasonable to do this. After all, they are letting the data inform the model choice and aren't using strong subjective criteria to define which models are in consideration. On the other hand, how many times have you wandered into a research problem and said, "I have 493 different working hypotheses that I wish to test"? And if you did, how much data do you think you would need to distinguish among them?

The problem here is that the scientists were awash in models. Consequently, the best fitting model or models could easily emerge through random chance alone. Allowing a priori a large number of candidate models greatly inflates the chances that the inferences won't be robust (meaning that a replicate study would find a different answer). The solution is relatively straightforward: think carefully about what you are hoping to test. What questions are you truly trying to answer? What are the main categories of hypotheses? Once you do this, you can get the list of candidate hypotheses near a dozen or so, a reasonable number.

In this case, the authors' intent was good. They wanted to make sure they were accounting for potential environmental factors that might confound their comparison. And they didn't know which environmental factors would be important. The problem lay in how they accounted for this lack of information. The honest approach would be to throw all of those variables into one model, call that the "environment" model, and then compare that to a model that only contained the conservation area effect, and to a third model that contained the conservation area effect plus all of the environmental factors. That approach would penalize the analysis for not knowing which environmental factors were important (and would likely provide incentive to the analysts to give some a priori thought as to which factors should be included, as each one included carries a potential AIC penalty if it doesn't improve the negative log-likelihood).

Advanced: Refinements of the AIC

If you start using the AIC regularly, you should be aware of modifications to the basic equation that are used to reduce bias. The original derivation assumed very large sample sizes. When sample sizes are small, the AIC has a bias towards favoring more complex models. We commonly use a version of the AIC that is corrected for small sample size, called AIC_c:

$$AIC_c = 2\left(-L_{max} + K + \frac{K(K+1)}{n-K-1}\right) \tag{9.7}$$

Notice that the bias adjustment term on the right increases as the ratio of K (the number of estimated parameters) to n (the sample size) increases.

continued

Another source of bias is overdispersion: the AIC is biased towards complex models when data are overdispersed. The QAIC adjusts the AIC to account for overdispersion in the following way:

$$QAIC = 2\left(\frac{-L_{max}}{\hat{c}} + K\right) \tag{9.8}$$

where $\hat{c}$ is the variance inflation factor which can be calculated in a number of different ways (Burnham and Anderson 1998).

Finally, you may see the term $QAIC_c$, which stands for quasi-AIC corrected for small sample size, which corrects for both small sample size and overdispersion.

Summary

- We often select models based on expected predictive performance.
- The AIC is an estimate of the expected ability to predict future observations.
- AIC and ΔAIC give a continuous measure of "degree of support by the data," and they are not meant to be used thoughtlessly via arbitrary cutoffs.
- Pay attention to the relationships between alternative models (e.g., nested, not nested), model fit, and overdispersion.

Exercises

1. Use model selection to ask whether geoduck observations as in section 8.5 are best described by a Poisson or a negative binomial distribution. Although it's not obvious from the math, the negative binomial is essentially a more complex version of the Poisson (as the parameter k goes to infinity, the negative binomial has the same variance as the Poisson).

2. Examine some journal articles that you have on hand, and see if they used the AIC for model selection. If they did, review their alternative models and determine which, if any, are nested within each other. Did the authors use the AIC sensibly in these cases to describe how the data supported their models?

10

Bayesian Statistics

10.1 Introduction

The statistical world is divided on approaches and philosophies used to make scientific inference. On one side, there are the frequentists. Frequentists believe that we can judge hypotheses by how likely the outcomes are. In other words, if my hypothesis were true, how frequently would I expect to see a result like the one I just observed? On the other side are the Bayesians. They assign degrees of belief to particular hypotheses, that is, they like to judge hypotheses by stating how likely the hypotheses are.

10.1.1 Are you a frequentist or a Bayesian?

Fortunately, there is a behavioral assay that will 100% correctly identify you as frequentist or Bayesian. Answer the following two questions to find out which you are.

First, imagine that I am in front of you, and I'm about to flip a coin to determine the outcome of some bet between us.

Question 1: What is the probability that it comes up heads?
Here we go—I put a blindfold on you so you can't see what happens. Now I'm flipping the coin (pause while it flies in the air and then lands on the ground).

Question 2: What is the probability that it came up heads?
If your answer to question 2 was the same as your answer to question 1, then congratulations—you are a Bayesian! If your answer to question 2 was different than that for question 1, congratulations—you are a frequentist!

10.1.2 Wait, what are you talking about?

In the frequentist view of probability, you can only assign probabilities to future events. So, before I flip the coin, there is a 50% chance that it will come up heads. Once I flip the coin, it either came up heads or did not. You don't know which it was, because I blindfolded you. A frequentist therefore says that the probability that it came up heads is either 0 or 1; it either did or did not.

Introduction to Quantitative Ecology: Mathematical and Statistical Modelling for Beginners. Timothy E. Essington, Oxford University Press. © Timothy E. Essington 2021.
DOI: 10.1093/oso/9780192843470.003.0010

Bayesian statistics uses a different interpretation of the word "probability." Instead of referring to the probability of future events, probability means *degree of belief*. In this case, you know full well that it was either heads or tails. But because you don't know which it was, you assign some degree of belief that it came up heads.

This distinction underlies all of the differences between the two methods. For example, what is the definition of a p-value (a classic frequentist concept)? Is it the probability that the null hypothesis is true? No, because, under frequentist statistics, the null hypothesis either is or is not true. You just don't know which one, just like our fictional coin toss above. As a result, you need to invoke all sorts of careful language to describe a p-value: it is the probability of getting the observed outcome, or a more extreme outcome, if the null hypothesis were true. What is the definition of a 95% confidence interval? Is it an interval that has a 95% chance of including the correct parameter value? No, because the interval either does or does not include the correct parameter value. Under the frequentist interpretation of probability, you define a confidence interval as a range of numbers that was calculated using a method that will include the correct parameter value 95% of the time.

In contrast, Bayesian statisticians go right ahead and assign probabilities—degrees of belief—to alternative hypotheses or to particular parameter values. Using Bayesian statistics, you can calculate the probability that a hypothesis was true, given the data. Bayesian statistics can also generate the equivalent of confidence intervals, usually called credibility intervals, that are interpreted as "having a 95% chance of including the true parameter value."

10.1.3   So, how is this different from likelihood?

We know that the likelihood of a model, or of a set of parameters, is defined as: $L(\text{model}) = P(\text{data} | \text{model})$.

In Bayesian statistics, we want to know the probability of the model, given the data: $P(\text{model} | \text{data})$. This means we need some way to relate $P(\text{data} | \text{model})$ to the $P(\text{model} | \text{data})$.

Derivation of Bayes' theorem

Bayes' theorem is names after Rev. Thomas Bayes , who wrote an essay in 1783 that included the relationship between two different conditional probabilities.

Conditional probabilities are of the form $P(A|B)$ or $P(B|A)$, which is the probability of the first event, given that the second event happened.

Recall some of the probability rules, particular those that tell us the probability of event A and event B happening:

$$P(A \text{ and } B) = P(A) \times P(B|A)$$

continued

We can write this same expression this way:

$$P(A \text{ and } B) = P(B) \times P(A|B)$$

The left-hand sides of these two expressions are the same, meaning that the right-hand sides are also equal to each other:

$$P(A) \times P(B|A) = P(B) \times P(A|B)$$

So that we can write either of the conditional probabilities as:

$$P(B|A) = \frac{P(B)P(A|B)}{P(A)}$$

and

$$P(A|B) = \frac{P(A)P(B|A)}{P(B)}$$

Bayes' theorem is therefore a tool to relate one conditional probability to another conditional probability.

10.2 What is Bayes' theorem, and how is it used in statistics and model selection?

Bayes' theorem simply relates two conditional probabilities to each other. Consider two events, A and B. We want to know the probability of A occuring, given that B has occurred (put another way, what is the probability that A is true, given that B is true). Bayes' theorem states

$$P(A|B) = \frac{P(A)P(A|B)}{P(B)} \qquad (10.1)$$

Suppose θ is either a model or a particular parameter value. We want to calculate the $P(\theta|\text{data})$, that is, what is the probability that the θ is true, given our data. We can simply substitute "θ" and "data" into the equation above, using "θ" for A, and "data" for B:

$$P(\theta|\text{data}) = \frac{P(\theta)P(\text{data}|\theta)}{P(\text{data})} \qquad (10.2)$$

The left-hand side is the thing we want to make some inference about, by assigning a degree of belief. The right-hand side is more complex, so we need to work through it:

1. We have $P(\theta)$. This is called the "prior probability," or the prior degree of belief that the hypothesis is true.
2. We have $P(\text{data}|\theta)$, which we already know how to calculate: this is just likelihood.
3. We have the denominator, which is the probability of getting the data.

The probability of getting the data for discrete alternative hypotheses is

$$P(\text{data}) = \Sigma_i P(\theta_i) P(\text{data}|\theta_i) \tag{10.3}$$

where each θ_i is a distinct hypothesis. In other words, if there are two hypotheses, the probability of the data equals the probability of the data if the first hypothesis is true times the prior probability that the first hypothesis is true, *plus* the probability of the data if the second hypothesis is true times the prior probability that the second hypothesis is true.

Because we call $P(\theta)$ the prior probability, we call $P(\theta|\text{data})$ the posterior probability. An easy way to think of this is that $P(\theta)$ is your degree of belief in the model before (prior to) you collect the data, $P(\theta|\text{data})$ is your degree of belief in the model after (posterior to) you collect your data.

10.2.1 Doesn't the prior probability influence the posterior probability?

It sure does. In fact, one of the main criticisms of Bayesian statistics is that it requires you to articulate some prior probability that your hypotheses are correct. Bayesian statisticians have a few counterarguments to this. The first is that we probably do have prior probabilities about hypotheses. Later we'll see an important example illustrating the problems associated with *not* acknowledging this. The second is that you can formalize the process of defining your prior probability based on past data and then update your posterior probability as new observations come in. This is called Bayesian learning, and it turns out your computer is doing this all the time. Search engines like Google, for instance, are always trying to anticipate what our internet search terms are going to be—it does this by accumulating information about us over time, so that each time we search is a new data point.

Bayesian statistics in a nutshell

What is Bayesian statistics?
It is a way of assigning a degree of belief to a particular model or hypothesis, given prior belief and new data. It relies on a different interpretation of "probability" as it applies to the true state of nature.

continued

What is the prior belief?
It is the degree of belief that you had, prior to collecting new information.

What are the benefits of Bayesian statistics?
Bayesian statistical methods are extremely useful in decision science, where the true state of nature is unknown but you have to make decisions about policy actions that account for this uncertainty. Much like an insurance company weighs risks and rewards to set costs of coverage, a conservation planner wants to make policy decisions that are most likely to be beneficial. Strict frequentist interpretation of probabilities don't allow you to approach this problem in a way that is intuitive, but Bayesian statistics is much more amenable to risk analysis. It is also possible to incorporate prior information in your calculation.

What are the disadvantages of Bayesian statistics?
You need to specify a prior probability, which is both a curse and a blessing, as we'll soon see. Also, Bayesian methods for complex models are more difficult to solve, and there are few standards for what level of posterior probability constitutes evidence for or against a hypothesis. Model diagnostics and model selection is also more complex.

10.3   Practice example: The prosecutor's fallacy

You are on trial for committing some crime. At the crime scene, an object was sampled for DNA, and a particular sequence was matched to your DNA. Moreover, this particular DNA sequence is fairly rare, only occurring in one out of every 1,000 individuals. During the trial, the prosecutor repeatedly mentions that "there is only a one in 1,000 chance that this person is innocent." Is the prosecutor correct?

What the prosecutor is saying is that $P(\text{innocent} \,|\, + \text{DNA match})$ is 0.001, based on the fact that the $P(+ \text{DNA match} \,|\, \text{innocent})$ is 0.001. Side note: the prosecutor may not be malicious; our brains are terrible at interpreting probability (Tversky and Kahneman 1983)). We know from Bayes' theorem that these two things are not identical, so we need to work through all of the pieces of Bayes' theorem to calculate the probability that you are innocent from the probability of getting a positive DNA match if you were innocent.

We want to calculate the $P(\text{guilty} \,|\, + \text{DNA match})$. Applying Bayes' theorem:

$$P(\text{guilty}| + \text{DNA match}) = \frac{P(\text{guilty})P(+\text{DNA match}|\text{guilty})}{P(+\text{DNA match})} \qquad (10.4)$$

We need to define $P(\text{guilty})$, the prior probability that you are guilty, if we want to know the posterior probability that you are guilty. In other words, before the DNA screening, what was our degree of belief that you were the culprit? This number will depend on a lot of things. Suppose your DNA was tested as part of a routine sample and recorded in an database, and there was otherwise no other obvious linkage between you and the crime. A reasonable P(guilty) in this case would be 1 divided by the total number of all possible suspects—there truly was no linkage between you and the victim; this might be 1 divided by the number of people in the entire community. Perhaps, instead, we could narrow that down to the number of people that had access to the crime scene. Suppose for the sake of illustration that only you and 299 other people had access to the crime scene. If that were the case, then P(guilty) would be $1/300 = 0.0033$.

We also need to know the P(+ DNA match $|$ guilty), which turns out to be very simple because we presume the DNA test is highly accurate and always comes up positive if there is a true DNA match. This means P(+ DNA test $|$ guilty) equals 1.

Finally, we need to calculate the probability of a positive DNA match. This is the denominator of Bayes' theorem. There are two ways of getting a positive match. You could really be guilty, in which case the probability of getting a positive DNA match is P(guilty) times P(+DNA match $|$ guilty). Or, you could be innocent and you received a false positive, in which case the probability of getting a positive DNA test is P(innocent) times P(+ DNA match $|$ innocent). We know that in probability statements, "or" means you add the probabilities together.

To calculate the posterior, we need to do a few final things. First, what is the P(innocent)? This must be equal to $1 - P$(guilty), or $299/300$, which equals 0.99667. What is the P(+ DNA test $|$ innocent)? This is the false positive rate—the rate touted by the prosecutor to argue for your guilt—which equals $1/1,000 = 0.001$.

Putting it all together, we find

$$P(\text{guilty}| + \text{DNA match}) = \frac{P(\text{guilty})P(+\text{DNA match}|\text{guilty})}{P(+\text{DNA match})}$$
$$= \frac{0.0033 \times 1}{0.0033 \times 1 + 0.99667 \times 0.001}$$

This equals 0.77, meaning that there is a 77% chance that you are guilty. It's surely not good news, but not nearly so dire as thinking that there is a 99.99% chance that you are guilty, as the prosecutor implied!

This outcome might be surprising, but it reveals the importance of the prior probability. Suppose instead of being a random sample in a DNA database, you were highly connected to the crime scene and victims in many ways. Instead of saying that the prior probability of being guilty is $1/300$, we might instead say it is $1/10$. If you use this prior, the posterior probability that you are guilty is 0.91, or 91%.

10.4   The prior

Any Bayesian analysis requires you to specify a prior probability for a model. On one hand, this is a limitation of Bayesian statistics, because there are often no formal processes for setting prior probabilities. On the other hand, we often do have prior information that is relevant for evaluating the outcome of experiments. Failing to include that information can lead to conclusions that are at odds with the accumulated evidence for or against a given hypothesis

10.4.1 Example: Do people have extrasensory perception?

In 2011, Daryl Bem, a noted social psychologist, published a paper called "Feeling the Future: Experimental Evidence for Anomalous Retroactive Influences on Cognition and Affect" (Bem 2011). In it, he described the results of several experiments that he claimed provided evidence that people could predict the unknown (extrasensory perception [ESP]). One of his experiments was as follows:

> This is an experiment that tests for ESP. It takes about 20 minutes and is run completely by computer. First you will answer a couple of brief questions. Then, on each trial of the experiment, pictures of two curtains will appear on the screen side by side. One of them has a picture behind it; the other has a blank wall behind it. Your task is to click on the curtain that you feel has the picture behind it. The curtain will then open, permitting you to see if you selected the correct curtain. There will be 36 trials in all.
>
> Several of the pictures contain explicit erotic images (e.g., couples engaged in nonviolent but explicit consensual sexual acts). If you object to seeing such images, you should not participate in this experiment.

There were 100 participants, each of whom chose between eight and twelve curtains. He found that, 53.1% of the time, participants selected the curtain containing an image when the photo was "erotic." In contrast, people chose the curtain with an image less frequently when the image was "neutral" (49.6%), "negative" (51.3%), "positive" (49.4%), or "romantic" (50.2%). He used standard statistical methods to show that the probability of getting these results, if the null hypothesis were true (i.e., no difference in ability to choose the photo across photo types) was 0.01. He therefore dismissed the null hypothesis and claimed that people had an enhanced ability to know which curtain contained a photo when the photo was "erotic."

This garnered plenty of media attention, and Bem even appeared on the TV show *The Colbert Report*.

10.4.2 Criticisms

There were many criticisms levied against this work. An obvious criticism is, "Why should only erotic photos elicit such capacity?" The other was a statistical one. Namely, extraordinary claims (such as the one Bem made) require extraordinary evidence. There have been hundreds of previous tests before this experiment to test for the same kind of "precognition." We want to know the probability that precognition exists and whether this experiment substantially altered the odds (Wagenmakers et al. 2011).

In a Bayesian framework, we want to know $P(\text{ESP}|\text{data}) = P(\text{ESP}) \, P(\text{data}|\text{ESP}) / P(\text{data})$.

Let's suppose, for the sake of illustration, that $P(\text{data}|\text{ESP}) = 0.50$. That is, the data have about a 50% chance of happening if there really is ESP. This is a pretty high number.

Even though the p-value isn't truly equal to $P(\text{data} \mid \text{no ESP})$, for illustration let's assume that is the case, so $P(\text{data} \mid \text{no ESP}) = 0.01$. To finish the calculation, we need to know $P(\text{ESP})$—this is used in the numerator, and again in the denominator in calculating $P(\text{data})$.

What do you think $P(\text{ESP})$ should be? Remember, this is the degree of belief that there is ESP (specifically, an precognitive ability that applies only to erotic photos) before the experiment was conducted. Personally, I would put this at a pretty low number, say, one in a million odds. Wagenmakers et al. (2011) put it at 10^{-20}!

Let's work through the calculations:

$$P(\text{ESP} \mid \text{data}) = 10^{-6} \times 0.5 / (0.5 \times 10^{-6} + 0.01 \times 0.999999) = 5 \cdot 10^{-5}.$$

This means the experiment increased the odds that there is ESP by about 50-fold (from 10^{-6} to $50 \cdot 10^{-5}$). Still, the odds are minuscule, and certainly not worthy of extensive press coverage!

10.5   Bayesian parameter estimation

So far, we have applied Bayesian analysis to test alternative hypotheses. The steps are (1) formulate your prior, (2) calculate the likelihood of the data, given the hypothesis, and (3) calculate the probability of getting the data by summing over all alternative models.

What if we want to assign probabilities to parameter values? If the parameter values can only take discrete values, we can apply the same method as we did for testing alternative hypothesis. Earlier (section 10.3), we considered two discrete alternative hypotheses: you were either guilty or innocent. Because the alternatives were discrete, we could calculate the denominator of Bayes' theorem using equation 10.3.

But let's say you are interested in tracking the rate of population growth of a threatened or endangered species to determine its prospects for recovery. It would be very useful to be able to say, "There is a 70% chance that the population growth rate is positive." To make this sort of claim, we need to assign probabilities to particular ranges of parameter values.

The challenge comes from the fact that parameter values are very rarely discrete alternative values, as in the case with the alternative hypotheses that we've considered so far. For continuous alternative parameter values, the denominator of Bayes' theorem is

$$P(\text{data}) = \int \pi(\theta) f(\text{data}) \mid \theta) d\theta \tag{10.5}$$

where $\pi(\theta)$ is the probability density function for the prior probability, and $f(\theta \mid \text{data})$ is the likelihood. So, the denominator now becomes an integral and, in many cases, this integral has no mathematical solution

Making matters even trickier is that we have to integrate the posterior distribution to say anything about probabilities, because our posterior probability is also continuous. To estimate whether population growth rate is positive, you would also have to integrate the numerator of Bayes' theorem, so we are faced with the following expression to deal with:

$$P(r > 0|\text{data}) = \frac{\int_0^\infty \pi(r)f(\text{data}|r)dr}{\int \pi(r)f(\text{data}|r)dr} \tag{10.6}$$

Before you panic, rest assured that a lot of the Bayesian methods are designed to deal with the problem that equation 10.6 poses: how can we estimate these probabilities with integrals that might never be solvable?

10.5.1 How do we do that?

There are two main ways of tackling the problem. The first is to approximate the integral in some way. Approximating the integral simply involves saying, "Yes, I understand that the variable is continuous, but I'm going to pretend as though it takes discrete alternative values." Once we pretend that r can only take certain discrete values, we can use the same tools that we've already applied to calculate the posterior probability. More generally, we approximate an integral $\int f(x)dx$ with $\sum f(x)\Delta x$, where $f(x)$ is our prior probability density for any value of x. Placing this into equation 10.6, we get the following:

$$P(r > 0|\text{data}) = \frac{\sum_{r=0}^{r=r_{\max}} \pi(r)f(\text{data}|r)\Delta r}{\sum_{r=r_{\min}}^{r=r_{\max}} \pi(r)f(\text{data}|r)\Delta r} \tag{10.7}$$

where $r_{\min}$ and $r_{\max}$ are the smallest and largest values, respectively, we believe r can take. Notice that Δr appears in the numerator and denominator, so it cancels out.

In practice, assume that r can be any value between -0.5 and $+0.5$ in increments of 0.005. Now the steps are the same as our practice example; we just have a lot of alternative discrete models. The steps are as follows:

- calculate the prior probability density for each candidate value of r, $\pi(r)$
- calculate the likelihood for each candidate value of r, $f(\text{data}|r)$, and multiply by $\pi(r)$; that gives you the numerator
- calculate the denominator as the sum of $\pi(r)f(\text{data}|r)$ over all different values of r
- divide each numerator by the denominator for each candidate value of r; this gives $P(r|\text{data})$

This gives you the posterior probability for each candidate value of r in an approximate case where r only takes discrete values. To get the probability that r is greater than 0, we would add all $P(r = 0.0|\text{data}) + P(r = 0.005|\text{data}) + \ldots + P(r = 0.5|\text{data})$.

Obviously, the smaller the increments of r you use, the more accurate your approximation will be.

10.5.2 Monte Carlo methods

The approximation method works great when you have one or two parameters, but it becomes unwieldy once you have more than that. Generally, you'd want to use many alternative values of the parameter values to get better precision. Suppose for each parameter you needed 200 alternative parameter values. If you had two parameters, you'd have to calculate the probability of all different combinations of parameters, which would be 40,000 alternative parameter values. With three parameters, it becomes 8,000,000, and with four you're dealing with 1,600,000,000 different combinations (yes, over a billion!). At that point, the approximation method becomes inefficient and unwieldy.

Monte Carlo methods are more efficient ways of estimating the posterior in these cases. Section 15.3 shows that these routines are intended to explore parameter space by randomly generating parameter values. In Bayesian statistics, they are one of several of randomized algorithms *that simulate random draws from the posterior probability density*. The idea is that direct calculation of the posterior is intractable, but, by using a Monte Carlo routine, we can simulate what many draws from the posterior probability distribution would look like. These samples reveal the posterior probability distribution.

We call these methods "Markov Chain Monte Carlo" (MCMC). The process begins by specifying (usually randomly) several different sets of parameter values. For simplicity, suppose you were estimating the mean and standard deviation of a normal distribution. You would have several points, each of which corresponded to a particular value for the mean and another value for the standard deviation. Each point becomes a "chain" through a series of random steps that are guided by the numerator of Bayes' theorem. You let these chains step through parameter space, and if they follow a particular set of rules, eventually they will start to resemble random draws from the posterior distribution (Gelman et al. 1995). There are many such set of rules, all of which are well beyond the scope of this text. Refer to a good Bayesian textbook such as those by Gelman et al. (1995); Gelman and Hill (2007), and Hobbs and Hooten (2015).

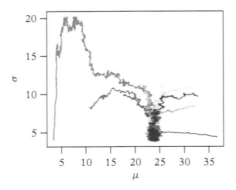

Figure 10.1 *MCMC chains converging on the posterior distribution.*

Figure 10.1 shows an example of ten MCMC chains, estimating the mean and standard deviation of a population from a sample of 100 data points. Each line and point depicts a chain. Note how they initially are all spread apart, but eventually they converge on a common area. Usually, we trim off the beginning of the chains, because they are still finding their way to where the posterior density is highest, and use the points that the chain visited later as random samples from the posterior density distribution. We can then look at the frequency distribution of posterior draws to generate the posterior distribution (figure 10.2). These are called the "marginal posterior distributions" because they are looking at one dimension (parameter) in what is a two-dimensional solution. Here the mode of the posterior probability distribution for μ is roughly 24, and μ might be anywhere between 22.5 and 25.5. The mode of the posterior probability for σ is centered near 4.5, and σ might be anywhere between 3 and about 6.5.

One problem is that the algorithm isn't guaranteed to converge (figure 10.3). Nine of the chains are doing the right thing, but then there is one chain up in the upper left-hand corner that hasn't figured out that it should be in the lower right-hand corner. If we didn't take a close look at this plot and generated the plots of the output, we'd see something strange (figure 10.4).

There are several diagnostics that one can do to test for convergence. Yet, while all of them can tell you with certainty that your MCMC did *not* converge, none of them can

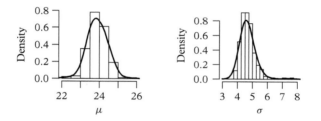

Figure 10.2 *Histograms of MCMC output, with fits drawn as lines.*

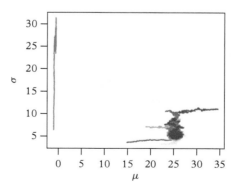

Figure 10.3 *The orange MCMC chain failed to converge on the posterior distribution.*

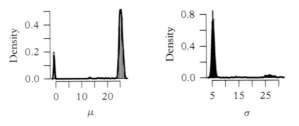

Figure 10.4 *Poor estimation of posterior distribution caused by lack of convergence.*

tell you with certainty that your MCMC *did* converge. In other words, the tests are for "lack of convergence," but it is possible that a MCMC routine can pass these tests, even though they did not converge. As a result, one always needs to be extremely careful using MCMC output and pass it through as many tests as possible.

10.5.2.1 Software to do MCMC

As of this writing, there are several options for doing MCMC on Bayesian models. Of course, you could code in your own routine (as I did above), but chances are these will not perform as well as algorithms that others have written. Some common ones are BUGS (Bayesian Inference Using Gibbs Sampling), JAGS (Just Another Gibbs Sampler), and Stan (which isn't an acronym at all). Stan is leading the pack at present, as it uses routines that sample the posterior better, has nice diagnostic features, and is much faster than the others (Monnahan et al. 2017).

10.5.2.2 Key points

- Monte Carlo methods are used to "simulate" draws from the posterior distribution.
- They are another way to solve the Bayesian integration problem.
- They are most useful for models with many parameters (more than three or four).

10.5.3 Laplace approximation

There is one final way to estimate the posterior probability distribution, using a method called the "Laplace approximation." Unlike MCMC, it does not simulate random draws but rather provides a different way of approximating the posterior probability distribution that works even when you have several parameters. The details are beyond the scope of the book, but essentially it finds the parameter combinations that maximize the numerator of Bayes' thereom $\pi(\theta) f(\text{data}|\theta)$ and then evaluates the curvature of that peak (does the numerator drop off sharply with changes in θ or more slowly?). Based on this, it approximates the shape of the full function $\pi(\theta) f(\text{data}|\theta)$, using a mathematical equation whose integral can be calculated. Programs such as INLA (integrated nested Laplace approximation) use this method. One advantage of this method is that it requires far less computational time than MCMC methods need.

10.5.4   Mechanics of using the prior

Bayesian analysis requires that we specify our prior belief that the parameter(s) of interest take certain values. We can do this by applying a probability density functions to quantify our prior degree of belief in the parameters. We need to complete two steps.

First, choose a probability density function that makes sense for the parameter that we are using. Generally, these will be continuous probability density functions, because our parameters are continuous.

Second, define parameters of those probability density function to accurately depict your prior degree of belief about the parameter that you are estimating. The parameters that describe the probability density function of our prior are called "hyperparameters," because they are not estimated and they dictate the prior probability of the parameters that are estimated.

10.5.4.1   *Example: The beta distribution*

Recall our coin flip example from section 9.1, where a coin was flipped ten times and came up heads eight times. We used maximum likelihood estimation to evaluate the evidence that the coin was fair: that the true value of the binomial parameter p was between 0.45 and 0.55. How would we address this in a Bayesian context with a prior belief?

First, we need to specify the probability density function for p. The beta probability density function is a good choice, because it describes the probability of a random variable that can only take values between 0 and 1 (section 7.7). This probability density function has two parameters, α and β, that together determine the mean and variance of p: α and β are the hyperparameters of the prior distribution of p, the parameter we are estimating. The mean of the beta distribution is $\alpha/(\alpha+\beta)$, and the entire beta distribution looks like figure 10.5.

We want to define the hyperparameters such that the prior accurately represents our prior belief. For instance, suppose we want a prior probability density such that the prior

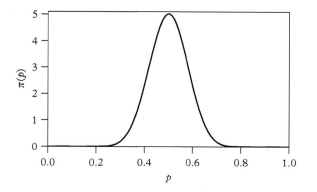

Figure 10.5 *Using the beta distribution to define the prior probability of p.*

probability that the coin is fair, that is, p is between 0.45 and 0.55, will be roughly 50%, that is, there are even odds that the coin is fair. The hyperparameters, $\alpha = 20$ and $\beta = 20$ (these are the values used to generate figure 10.5) roughly achieve this.

How did I know this? The easiest way is to use the difference in cumulative probability functions between $p = 0.55$ and $p = 0.45$ (the cumulative probability functions in either R or Excel give you the probability that a random variable is less than or equal to some value. By subtracting cumulative probabilities between two points, b and a, you get the probability that the random variable falls between a and b). Here, the area under the curve between 0.45 and 0.55 equals 0.47, or a 47% probability.

If you were to use this prior probability function to calculate the posterior probability, using the discrete alternative approximations, you would follow the steps above:

- Specify the alternative values of p. For instance, you might choose all values of p between 0.005 and 0.995 in increments of 0.005.
- For each p, calculate the prior probability density from the beta distribution with hyperparameters $\alpha = 20$ and $\beta = 20$.
- For each p, calculate the likelihood.
- Multiply the prior probability and likelihood for each value of p. This is the numerator of Bayes' theorem.
- Calculate the probability of the data by summing the numerator of Bayes' theorem over all of the alternative values of p.
- Divide the numerator of Bayes' theorem for each value of p by the probability of the data.

10.6   Final thoughts on Bayesian approaches

We introduced this part of the book by claiming that it is useful to be able to compare multiple hypotheses directly. Yet, I haven't said anything yet about Bayesian model selection in an actual ecological context. There is good reason for that: Bayesian model selection is a rapidly evolving field, and all methods require a fair amount of understanding of Bayesian methods (Hooten and Hobbs 2015). To the extent that you can pose your alternative models as alternative values of parameters, Bayesian parameter estimation can help you assign probabilities to those hypotheses. This was what we did in our brief example of estimating population growth rate to assign a probability to the belief that population was growing.

I won't answer the question "Should ecologists become Bayesian?" (Dennis 1996). I admit I'm agnostic. Worse still, I'm an opportunistic pragmatic agnostic. I use frequentist and information-theoretic methods when I can, simply because the calculations are easier, and I've never had a reviewer criticize me for failing to justify the choice of frequentist statistics. Yet, some models become too difficult to fit using the numerical methods described in the skills section. In these cases, I take advantage of the fact that Bayesians have worked out numerical methods that will work in a wide range of models.

Other times, I've wanted to propagate uncertainty in a series of model calculations, and Bayesian estimation is simply an easier way to do that. Finally, I have found Bayesian methods particularly useful when I've had multiple data sets and I want to use both to estimate some property (Essington et al. 2021). Here I fit model to one data source with "noninformative priors" and then used the resulting posterior probabilities as prior probabilities for a second data source.

Part III

Skills

This part of the book is intended to give more detailed instruction on the technical aspects of modeling. Many presume that mathematical and statistical modeling must be done in programming languages that can expand to accommodate the complexity of models and model analyses that we often face in quantitative ecology. To a certain extent, that is true. If you are embarking on a career in quantitative ecology, then why not gain familiarity with the technical tools that are used in that field? On the other hand, what if you do not know whether you intend to use these skills on a regular basis? Or what if the idea of learning an unfamiliar programming language is overwhelming given that you are grappling with plenty of new challenging material to master? Does that mean that ecological modeling is not for you, that there is no way to learn the concepts that matter? Of course not. In fact, I'd argue that many people can better learn how to model by starting with spreadsheets and then expanding from there.

This section will provide guidance on using both spreadsheets and R (a commonly used programming language and environment). You can decide which is best for you. Below I summarize what I view as the main advantages and disadvantages of each.

Why would I use spreadsheets?

You may be reasonably familiar with spreadsheets, know how to write a formula, know the difference between absolute and relative cell references, and be reasonably crafty in using spreadsheets to tackle your tasks. You might find the visual nature of spreadsheets useful. You may want to practice the exploration of ecological models through relatively tractable examples to deepen your understanding of the main underlying concepts. You are not particularly interested at this point in applying these skills to your research problems, and are therefore not too concerned with reproducibility and the ability of others to understand your calculations. You are willing to put up with the syntax of spreadsheet formulas (and how they can be hard to debug) in exchange for the other comforts and uses described above.

Why would I use a programming language?

You don't find yourself particularly more comfortable in a spreadsheet environment over a programming language. You want syntax that is easy to read and to debug. You don't care if you don't always see the model output, or the intermediate calculations therein. You want to have few limitations on the model's complexity, or on the complexity of model analyses that you might one day perform. You are certain that you will be applying these methods for your research and want to get a jump start on the technical skills that will grow as you continue to gain more experience. You want to be able to take advantage of vast resources of user-created functions and packages that perform sophisticated tasks for you without having to build them from the ground-up. You want to have easily sharable and reproducible quantitative analyses to support your research.

In my experience, most people are more comfortable *learning* about ecological models in spreadsheets than in programming languages. Individual results may vary.

Okay, I want to use a programming language. But what is R and why should I use that one?

R is an free and open-source language and software environment (http://www.r-project.org/). In recent years, it has become one of the most widely used tools for all manner of ecological modeling, though it is in statistical modeling that R really outpaces others. It is also capable of producing outstanding graphics (in fact, R was used to create nearly all of the graphics in this book).

For those reasons, R is a great place to start. By learning R—as opposed to other languages such as C++, Fortran, Matlab, Maple, and Mathematica—you will be able to freely and easily share your code with a wide community and take advantage of the multitude of user-generated packages.

That said, R is just one option. Will R be the leading language in twenty-five years? Is R necessarily the "go to" environment in your subdiscipline? Is R better at all things all of the time? Of course not. Even now, Python (another free, open-source programming language: http://www.python.org/) is used almost as often as R. Still, by learning R, you'll learn the programming basics that are widely shared across platforms. The details of the syntax will be different, but the logic and structure are often fairly similar.

11

Mathematics Refresher

Here I review some common mathematical functions, derivatives, and matrix operations that you might find useful. They are presented in no particular order, with the assumption that this section will largely serve as a reference once you are digging into a model.

11.1 Logarithms

You probably think of logarithms as being of two kinds: "regular" logarithms, or base 10, and "natural" logarithms, or base e. In this book, I use the notation log to refer to base e logarithms. Why? Because that is what mathematicians do (and most programming languages). So, we might as well follow suit.

The notation $\log(x)$ is the natural logarithm (or $\log_e$ of x) and $\log_{10}(x)$ is the base 10 log of x; $\log(x)$ simply means find the power to which the base value must be raised to produce the value x. So, $\log_{10}(100)$ equals 2, because 10^2 equals 100. Logarithms are undefined at 0, and $\log(1)$ equals 0.

I will sometimes use the notation $\exp(x)$ to replace e^x. In other words, $\exp(x)$ means take Euler's number e and raise it to the power of x.

11.1.1 Common operations with logarithms

$$ab = e^{\log(a)+\log(b)}$$
$$\log(ab) = \log(a) + \log(b)$$
$$\frac{a}{b} = e^{\log(a)-\log(b)}$$
$$\log\left(\frac{a}{b}\right) = \log(a) - \log(b)$$

Introduction to Quantitative Ecology: Mathematical and Statistical Modelling for Beginners. Timothy E. Essington,
Oxford University Press. © Timothy E. Essington 2021.
DOI: 10.1093/oso/9780192843470.003.0011

11.2   Derivatives and integrals

In the following equations, a is a constant, x and y are variables, and $f(x)$ and $g(x)$ are functions:

$$\frac{da}{dx} = 0$$

$$\frac{dax}{dx} = a \text{ or, more generally, } \frac{dax^n}{dx} = nax^{n-1}$$

$$\frac{d\log(x)}{dx} = \frac{1}{x}$$

$$\frac{d(f(x) + g(x))}{dx} = \frac{df(x)}{dx} + \frac{dg(x)}{dx}$$

The last equation is maybe the most useful rule (often called the sum rule). It means you can break down any derivative into the component functions that are added together, take the derivative of each, and add them all together. For instance,

$$\frac{d(ax^2 + bx + c)}{dx} = \frac{d(ax^2)}{dx} + \frac{d(bx)}{dx} + \frac{d(c)}{dx}$$

$$= 2x + b + 0$$

The product rule: the derivative of two functions multiplied together, is

$$\frac{df(x)g(x)}{dx} = g(x)\frac{df(x)}{dx} + f(x)\frac{dg(x)}{dx}$$

The quotient rule: the derivative of one function divided by another, is

$$\frac{d\frac{f(x)}{g(x)}}{dx} = \frac{\frac{dg(x)}{dx}f(x) - \frac{df(x)}{dx}g(x)}{g(x)^2}$$

Finally, the chain rule is as follows: when a variable z is a function of y, and y is a function of x, then we can find the derivative of z with respect to x as

$$\frac{dz}{dx} = \frac{dz}{dy} \cdot \frac{dy}{dx}$$

You'll never need to compute an integral in this book but it is useful to know what an integral means. An integral is an "anti-derivative," but is more commonly thought of as a way to find the "area under the curve." That is, given some function $f(x)$, and some bounds a and b, you wish to find the area between the function $f(x)$ and the origin, 0, between $x = a$ and $x = b$. We denote this with

$$\int_a^b f(x)dx$$

11.3　Matrix operations

11.3.1　Dimensions of matrices

Any matrix, here denoted **A**, has a specified number of rows and columns. We often indicate this using the following notation:

$$\underset{m \times n}{\mathbf{A}}$$

where m is the number of rows, and n is the number of columns.

11.3.2　Adding two matrices

If you want to add two matrices, **A** and **B**, first you have to confirm that they have the same dimensions. If this is true, then

$$\mathbf{C} = \mathbf{A} + \mathbf{B} = \begin{vmatrix} a_{11} & a_{21} \\ a_{12} & a_{22} \end{vmatrix} + \begin{vmatrix} b_{11} & b_{21} \\ b_{12} & b_{22} \end{vmatrix} = \begin{vmatrix} a_{11} + b_{11} & a_{21} + b_{21} \\ a_{12} + b_{12} & a_{22} + b_{22} \end{vmatrix}$$

11.3.3　Multiplying two matrices

Matrices can only be multiplied if the number of columns in the first matrix equals the number of rows in the second matrix. If you multiply a $m \times n$ (m rows, n columns) matrix by an $n \times k$ matrix, the resulting product will have m rows and k columns:

$$\mathbf{C} = \underset{3 \times 2}{\mathbf{A}}\ \underset{2 \times 1}{\mathbf{B}} \tag{11.1}$$

$$= \begin{vmatrix} a_{11} & a_{21} \\ a_{12} & a_{22} \\ a_{13} & a_{23} \end{vmatrix} \begin{vmatrix} b_{11} \\ b_{12} \end{vmatrix} \tag{11.2}$$

$$= \underset{3 \times 1}{\begin{vmatrix} a_{11}b_{11} + a_{21}b_{12} \\ a_{12}b_{11} + a_{22}b_{12} \\ a_{13}b_{11} + a_{23}b_{12} \end{vmatrix}} \tag{11.3}$$

The transpose of a matrix **A** simply switches the rows and columns of a matrix:

$$\mathbf{A} = \begin{vmatrix} a_{11} & a_{12} \\ a_{21} & a_{22} \end{vmatrix}$$

$$\mathbf{A}^T = \begin{vmatrix} a_{11} & a_{21} \\ a_{12} & a_{22} \end{vmatrix}$$

11.3.3.1 The identity matrix

The identity matrix is an $m \times m$ matrix with 1s on the diagonals and 0s elsewhere. It works in a equivalent manner as multiplying any value by 1. That is, any matrix multiplied by the identity matrix will return the original matrix:

$$\underset{4\times4}{\mathbf{I}} \times \underset{4\times2}{\mathbf{A}} = \underset{4\times2}{\mathbf{A}}$$

12

Modeling in Spreadsheets

You can do a lot more in spreadsheets than you probably think you can. I regularly use spreadsheets when I want to understand how a part of my model works or when I want to quickly create a simple model to answer a specific question.

Before you proceed, you should know the basics of using spreadsheets, understand the difference between absolute and relative cell references, and be able to write formulas with both of these. You should also know the basic spreadsheet functions like sum, average, count, and so on. Finally, you should be able to make simple plots (like x-y scatter plots).

Three commandments for modeling in spreadsheets

1. Put your parameters up top
Your life will be much easier if you first identify all of the parameters in your model and use cells at the top of the spreadsheet (or anywhere that is easy to find) to specify the parameter values. Then, in the model calculations, simply refer to the cells that contain the parameter values. If you embed parameter values in your equations, it will be nearly impossible to debug and will really restrict the capability of the spreadsheet.

2. Organize information in columns
If you have lots of columns but only a few rows, you end up with a messy, hard-to-follow spreadsheet. So, if you are doing a simulation model that simulates some state variables over time, have each row in the spreadsheet represent each time point, and the columns be different state variables (or different values needed for calculations at each time step).

3. Keep it simple
A complex, disorganized spreadsheet will probably lead to a disorganized model analysis. Start off by making each column's calculation very simple and then grow your model with additional columns as needed. Occasionally clean up the work space as you continue to develop your model (e.g., remove unnecessary graphs that are cluttering up the spreadsheet).

Introduction to Quantitative Ecology: Mathematical and Statistical Modelling for Beginners. Timothy E. Essington, Oxford University Press. © Timothy E. Essington 2021.
DOI: 10.1093/oso/9780192843470.003.0012

12.1   Practicum: A logistic population model in Excel

The best way to see the functionality of spreadsheets is through an example. Here you will code up a logistic population model, projecting population sizes twenty years from now, given an initial population size, a maximum population growth rate, and carrying capacity.

First, dedicate a portion of the spreadsheet to specify the model parameters and their values. Usually, we put these right at the top of our spreadsheet. Often, we'll type in the parameter names in one column, and parameter values in the adjacent column. Suppose the initial population size is 10, the maximum growth rate is 0.2 yr^{-1}, and the carrying capacity is 100. At the top of your spreadsheet, in cells A1 through B3, you might have the something that looks like table 12.1.

In this way, cell B1 contains the parameter value for the initial population size N_0, B2 contains the parameter value for maximum growth rate r, and B3 contains the parameter value for the carrying capacity K.

Always put your parameters and parameter values in an prominent place on your spreadsheet. That way it is easy to change the parameter values and explore how that affects the model behavior.

Now you need to calculate N_{t+1} based on N_t, following section 2.6, starting in the initial year (here called year = 0), all the way out to year = 20. To begin, choose a row below our parameters, say, row = 5, and in column A type in "Year" and in column B type in "Nt". These are just labels to make it easier to remember what each column is doing.

In the A column, starting in row 6 and continuing to row 26, enter in 0, 1, 2,..., 20 to denote the time step that each row depicts. Alternatively, you can enter 0 in cell A6, and, in cell A7, type in the formula = **A6 +1** and then copy that formula into cells A8 through A26.

The next step is to enter the model equations into the spreadsheets in column B. Cell B6 contains the initial population size. In this cell, enter the equation = **B1**. This says to take the value that resides in cell B1 and return it into this cell. Now, in cell B7, use the recursive equation, referring to the population size for the previous time step to get the population size for the current time step. The function would be:

= B6 + B6* $B$2* (1 - B6 /$B$3)

Table 12.1 *Putting your parameter names and values at the top of your spreadsheet*

	A	B
1	N_0	10
2	r_logistic	0.2
3	K	100

Table 12.2 *Abbreviated spreadsheet layout*

	A	B
1	N_0	10
2	r_logistic	0.2
3	K	100
4		
5	Year	Nt
6	0	=B1
7	1	=B6+ B6* $B$2* (1 - B6 /$B$3)

Note that we used absolute cell references to refer to the parameter values r and K. We do this so that we can simply copy this formula into cells B8 down to cell B26, to calculate population size for each time step.

Table 12.2 shows an abbreviated layout with spreadsheet functions.

The final step is to make an x-y scatter plot, with Year on the x axis and N_t on the Y axis to view the model behavior.

If you have multiple state variables (e.g., the competition model described in section 4.2), you would simply add more columns to accommodate them and then add parameters and their values as necessary.

This is a good time to illustrate one of the downsides of modeling in Excel. Take a look at the formula in cell B7. It is not pretty. You have all manner of cell references, some of which are absolute, and some are relative. It certainly works, but what if there was a bug in your spreadsheet and you wanted to make sure you entered the equation properly? Because the spreadsheet syntax is very different from how we write our model equations, it is harder to spot errors.

12.1.1 Naming spreadsheet cells

One way to improve the readability of spreadsheet formulas is to assign names to individual cells. That is, we can assign the name "N_0" to the cell that contains the initial population density. Then, instead of typing in the cell reference based on the row and column, we instead just type in the name of the cell.

Most spreadsheet software allows you to assign names to cells. The only problem is that each does so slightly differently, and what works on one operating system may not work on another. A simple way that seems to work in Excel regardless of the operating system used is to notice a small box in the upper left-hand corner of your spreadsheet (to the left of the formula bar) that shows the column letter and row number for whatever cell you have currently selected. If you select cell B1, this little box will show "B1." While B1 is still the active cell, click on this box such that the cell address becomes highlighted.

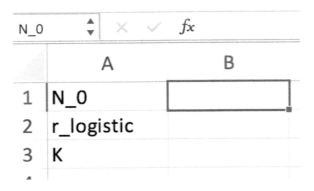

Figure 12.1 *The cell B1 has been named N_0, as indicated by the box in the upper left-hand corner.*

Now just type in the name you would like to give this cell. Because this cell contains the value for parameter N_0, type in **N_0** and hit Enter. When you are done, it should look like figure 12.1.

Once you have named the cell, you can reference it in a formula using **N_0** instead of the cell reference B1. Go ahead and name cell B2 **r_logistic**, and cell B3 **K**. Once that is done, your model formulas become much simpler. In cell B6, enter = **N_0**. And, in cell B7, enter =**B6 + B6*** **r_logistic*** (1- **B6** /**K**), and copy that formula down into cells B8 to B26. The syntax still isn't perfect, as it still needs to use the expression B6 to refer to population size in the previous time step. But is is much improved because at least the model parameters are written out in the same way as they are in our mathematical expressions.

There are a handful of constraints on the names you can assign to cells. For instance, you can not use "r," "c," or any name that is already used as an Excel function (e.g., "sum," "average"). That's why I used **r_logistic** instead of just **r** above. Also, you can't assign a name that resembles a cell reference. So, you can't name a cell something like "A1." That's the reason for the underscore in N_0 (though, technically, I could have gotten away with it, because there is no cell N0).

12.2   Useful spreadsheet functions

Below are some functions that often make modeling a bit easier. They are presented in no particular order.

=IF(condition, then, else)
This command says that if the "condition" is met, then do the operation listed under "then," or else do the operation listed under "else." For example, you might say =if(A10<0, 0, A10* B10, 0). This means that if A10 is less than zero, then return 0, and if A10 exceeds zero, return the value obtained by multiplying cells A10

and B10 together. You can nest multiple if statements together (e.g., if (condition,1 if(condition2,A,B),C)).

=COUNTIF(countrange, count criteria)

This counts the total number of entries in the specified range (countrange) that meets some specified condition (e.g., equal to some number or text string).

=SUMIF(range,criteria,sumrange)

This finds the rows within the first input range that meet the specified criteria (e.g. equal to some number or text string), and then sums the values in those same rows within a second range of numbers (sumrange).

=BINOM.DIST(x, trials, probability, cumulative) =POISSON.DIST(x, mean, cumulative) =NORM.DIST(x, mean, standard deviation, cumulative)

These are three common functions for returning the probability mass or probability density from an outcome, x. Note that they all include as the final argument **cumulative**, which you enter as either TRUE, if you want the function to return the probability of getting an outcome less than or equal to x, or FALSE, if you want the function to return the probability or probability density for x. Most of the time, you will probably use FALSE. Note that the built-in probability density mass function for the negative binomial uses a parameterization that is different from the one we use in ecology, so you can't easily use it.

=RAND()

This returns a uniform random variable between 0 and 1.

=BINOM.INV(trials, trial probability, probability value) =NORM.INV (probability value, mean, standard deviation)

These formulas return the value of a binomial or a normal distribution corresponding to a particular probability value. For example, a probability value of 0.5 will return the mean of the distribution. If you couple this with RAND(), you have a formula that will generate random numbers from each distribution:

=BINOM.INV(trials, trial probability, RAND()) =NORM.INV(RAND(), mean, standard deviation)

12.2.1 Exercise

Create a list of 100 random numbers drawn from a normal distribution with a mean of 100 and a standard deviation of 20.

Create a new spreadsheet, and name one cell "mean" and another cell "sd." Insert values 100 and 20, respectively, into these cells. Now, in another cell, use the NORM.INV command to create a random number drawn from the normal distribution:

=NORM.INV(RAND(), mean,sd)

Copy this cell, and paste it down the column so that you have 100 random numbers.

To confirm that this worked, use Excel to calculate the mean and the standard deviation of your random numbers. They should be very close to 100 and 20, respectively.

12.3   Array formulas

Spreadsheets allow you to do matrix algebra and other calculations that are functions of entire matrices and arrays. The secret to using these commands is to remember the following : Ctl-Shift-Enter. That is, rather than hitting Enter after typing in a formula, you hit Ctl-Shift-Enter to enter an array formula. You only get one try to get this right: if you accidentally hit Enter, you'll have to start all over.

One useful array formula is =FREQUENCY. This function will help you generate frequency histograms of data stored in a particular range. The syntax is

=FREQUENCY(data range, bins range)

where the bins range lists the "bins" that Excel uses to calculate the number of observations that fall within each "bin." Remember ... Ctl-Shift-Enter! (Or Cmd-Shift-Enter.)

12.3.1   Exercise

Create a frequency histogram of the 100 random numbers that you created above.

In a column somewhere on your worksheet, enter a list of numbers from 50, 60, 70, ... to 150. Now select the adjacent column, and enter the frequency formula in the formula bar: for the data range, list the range of cells containing the random numbers, and, for the bins range, enter the range of cells that list the numbers 50 to 150. Hit Ctl-Shift-Enter and you should have a frequency counts for each number in your bin range. Make a bar chart of this histogram, and confirm that it looks reasonably like the "bell-shaped" normal distribution.

12.4   The data table

Many times you'll use the entire spreadsheet to generate one number (say, log-likelihood from multiple observations) based on a number of parameters. Often you will want to know how this number varies depending on the values of the parameters. That means you have to manually enter in a value for the parameter, see what the model result is, write it down, and repeat this many times until you've tried all of the parameter values that you are interested in. This is slow but feasible if you have only one parameter that you're interested in. But if you are interested in combinations of two parameters, then you'll have blisters on your fingers before you finish. Fortunately, spreadsheets have automated this for us in the "data table" command. Here I'll give instructions that work in current versions of Excel.

12.4.1   Exercise

Calculate the effect of r on final population size.

Open your logistic population model created in section 12.1. Your model performs many calculations on a spreadsheet to predict population size at time 20, and this number is found in cell B26. Now, suppose you want to determine how year 20 population size depends on the value of r, for values of r ranging from 0.1 to 2 in intervals of 0.1.

Make a column of numbers ranging from 0.1 to 2 in intervals of 0.1 in cells D6 to D25. In cell D5, type in a formula that says return the value in cell B26 (e.g., **=B26**). It will look something like table 12.3. The values that appear in column D are the different values of r that we want to apply.

Select cells D5 through E25 (this is where the table will be inserted). PC users, select the "Data" menu in the ribbon bar. Under the category called "Data Tools," there is a command called "What-if Analysis." Select "What-if-Analysis," which will pop up a menu with three choices. Select the last one, "Data Table." Mac users, go directly to the "Data" main menu and select "Data table."

You'll see a pop-up window that asks you for the "row input cell" and the "column input cell." What this is really asking you is, "What parameter do the numbers in D6 to D25 refer to?" In this case, we don't have a row input cell (all of the parameter values are in a single column), so you can ignore that part. The column input cell is the cell that contains the value for parameter r (i.e., B2). Either type in the cell address, B2, or type in the parameter name, **r_logistic**.

Click "OK," and Excel will calculate population size for all values of r listed in the column E6 to E25.

Suppose you want to repeat this for different two different values of K, 100 and 120. Setup the cells in column D6 to D25 the same as before—these will be the values of r that Excel will use. Now, in cells E5 to F5 (or as many different columns as you need), type in the different values of K that you want to use. This means you have values of r specified in a column, and values of K specified in a row. Because this is a two-dimensional table, we'll put the value that you want calculated (population size: =B26) in the cell just *above* the column of r's and to the *left* of the row of K's, which in this case would be cell D5. Your layout should look like table 12.4.

Table 12.3 *Setting up your data table*

	D	E
5		=B26
6	0.1	
7	0.2	
...	...	
25	2	

Table 12.4 *Setup for a two-dimensional data table testing two values of K*

	D	E	F
5	=B26	100	120
6	0.1		
7	0.2		
...	...		
25	2		

Now select the entire range of cells (D5 through F25), go to "What-if analysis" and select "data table" where you'll get the same window as above. Under "row input cell", specify the cell on the spreadsheet that contains the value of K, and under "column input cell", specify the cell on the spreadsheet that contains the value of r. Click OK, and the entire data table will automatically fill itself out.

12.5 Programming in Visual Basic

Sometimes our models outgrow spreadsheets. We can use the Visual Basic script editor embedded in Excel to interact with our spreadsheets to help us out. Of course, if you are generating models that need a lot of scripted commands (e.g., programming), then the benefits of using spreadsheets start to disappear. If that is the case, perhaps it is time to start looking at R. But sometimes we only need a simple way to automate spreadsheet calculations, and Visual Basic helps us do that.

So, why is this useful? Suppose you have a spreadsheet that performs many calculations to return a single number. You want to repeat those calculations many times and save the output. For instance, above (section 12.3.1), you generated 100 draws of a normal distribution and then calculated the mean of those draws. Suppose you wanted to see the standard deviation of the means (in other words, the standard error). You would have Excel recalculate the spreadsheet many times, save the resulting sample mean for each, and place it on your spreadsheet. Imagine doing that 100 times! Instead, we can write a quick code to automate this type of calculation.

Let's try that. Open up the spreadsheet you used above (section 12.3.1) to estimate the sample mean from a random draw of 100 normal random variables. Presume that your sample mean is calculated in cell C1. You want your spreadsheet to calculate the sample mean 100 times and return each sample mean into column D, starting in row 2.

Visual Basic is included in your Excel software (usually), but you might need to activate it. On a PC, go to "Excel Preferences," select "Ribbon," and make sure that the "Developer Tab" is checked. Once you have done that, you can open the Visual Basic editor by clicking on the Developer Tab and then clicking on "Visual Basic Editor." On a

Mac, you can do the same, but I've found that I can get to Visual Basic through the main menu items \Tools\Macros\Visual Basic Editor. Once you do that, you'll get a screen that looks pretty useless at first. Don't worry, we'll find a use for it.

Within the Visual Basic editor, you will create a new *module* that contains your code. Go to the menu called "Insert" and select "Module." A blank page should open up. In this blank page, you can write commands and then call this series of commands from within Excel.

The first thing you need to do is to specify our code as a subroutine and give it a name. Let's call the routine "samplemeans." To do this, type in **Sub samplemeans()** on the first line of the module, then hit Enter. You'll notice that Visual Basic automatically places a **End Sub** at the bottom of the page. That just tells Visual Basic when the subroutine is finished. You will place all of the commands in between the first line and this **End Sub** line.

To make Excel repeat the spreadsheet calculations, we'll use a common programming tool called looping. Looping simply means that you are going to put down a series of commands and then tell Visual Basic to loop through those commands, repeating them for a specified number of times. We often use variables like "i" as counters for the loop. A common syntax looks something like this:

```
for i = 1 to 100
   do some commands
next i
```

This means that i is a counter variable that first takes the number 1. Visual Basic will run the commands listed in the brackets and, when those have been completed, will assign to i the next number in the sequence, in this case, 2. This will continue for as long as i is less than or equal to 100. Then the routine will stop.

In your case, your commands are pretty simple. You want them to do the following:

1. Recalculate the entire spreadsheet.
2. Retrieve the sample mean.
3. Place the sample mean in the the spreadsheet.

Given those needs, you code might look like this:

```
Sub samplemeans()

for i = 1 to 100
Sheet1.Calculate
xbar = range("C1")
Sheet1.Cells(i+1, 4) = xbar
Next i

End sub
```

The first line in the loop tells the spreadsheet to recalculate all of the formulas on Sheet1. The second line retrieves the value of the sample mean that resides in cell C1 (if your sample mean is elsewhere, put in the correct cell here). The last line is a little tricky. Here it is placing the output in the fourth column of the spreadsheet (i.e., column D). But, in each iteration, it uses a different row. For the first iteration of the loop, the results is placed in the $i + 1 = 2^{nd}$ row. For the second iteration of the loop, the answer is placed in the third row, and so on. In other words, by placing the $i+1$ for the row, it will fill the D column from row 2 to row 101 with the sample means.

Because the loop is set to go until i is 100, this will automate the process of repeating the random draws of numbers and calculating the mean 100 times.

I find it easiest to run the code from the spreadsheet itself. Move over the spreadsheet. In the Developer Tab, select "Macros." Here you should see "samplemeans" listed as an available macro. Simply select this, and click on "Run," and it should run.

You will probably get an error at some point. Debugging is a slow process, often made easier by "commenting" out lines of code. In Visual Basic, you can add a '(a single hanging apostrophe) on any line, and Visual Basic will ignore anything that comes after that. A common debugging strategy is to comment out lines, one or more at a time, until the program runs without returning an error. When that happens, it means that the error is likely (though not always!) in one of the lines that you commented out.

The important points are as follows:

- We can repeat calculations on a spreadsheet by using a loop.
- Visual Basic can both take information from a spreadsheet and place information into a spreadsheet.

One final note on writing macros in Visual Basic. If you try to save a spreadsheet with macros within them as a normal Excel file, it will remove the macro from the file. For that reason, always save as a "Excel Macro-Enabled Workbook."

12.5.1   Creating your own functions in Visual Basic

You can also create your own functions in Visual Basic and call them in your Excel spreadsheet. This is useful if you find yourself commonly needing to perform a complex calculation that requires a complex cell formula.

Just as you would when creating subroutines, start by going to Visual Basic, and select a new **module**. To create a function that takes two numbers, x and y, adds them together, and divides by x, you would enter this:

```
Function sumanddivide(x,y)
Application.Volatile
sumanddivide = (x + y) / x
End Function
```

The first line uses the call Function to tell Visual Basic that is is intended to be a function that you can call in an Excel spreadsheet. The second line tells Excel that it should treat this function like all other functions (i.e., it should recalculate whenever anything on the spreadsheet is changed). The third line has two important components. The right-hand side calculates the quantity that we want the function to return. The left-hand side has the function name (sumanddivide). This is how Visual Basic knows what value to return when the function is called in Excel.

Now, in your Excel spreadsheet, you can call this function:

=sumanddivide(3,4)

which will return 7 divided by 3 = 2.333.

13

Modeling in R

First, a disclaimer. This book section is not a substitute for a dedicated book or training program on using R. Rather, this is going to give the basics that will get you started. The internet is full of useful tutorials that cover the basics of using R. For that reason, I won't repeat them here. Just do an internet search for "R tutorial," "guide to using R," or "R for beginners."

Second, there is no deep mystery in using R. Unlike so-called compiled languages (C++, Fortran), you can run your scripts directly in the environment. In other words, you can quickly and easily run your code, view the output, and see if there are errors.

Third, the main thing to know is that R allows you to generate things called objects, which are saved in the computer's memory and upon which computations are applied. These objects will be your parameter values, state variables, functions, data, and so on. Everything that goes into R is an object. Consequently, there are a lot of different object types to handle this flexibility. For example, objects can be lists, they can be data frames, or they can be vectors. In fact, it can be a bit overwhelming at times, but fortunately you can make a lot of progress at first by ignoring these distinctions and only paying attention when you see error messages that refer to object types.

Finally, the R environment takes some getting used to. You can run everything in R via a command terminal or you can use the R graphical user interface, but many people I know use Rstudio. Rstudio is another free software that creates a useful environment. Here you'll have windows to view your code (your script), the command console where commands are entered and run, and ways to view the objects that you've created. It even includes a window to view plots that you've created.

A note on example code

Throughout this section, I usually will show example R code that is intended to be maximally understandable, not maximally efficient. What does that mean? Well, there may be a clever way to condense five lines of code into one, but, by doing so, the intermediate calculations and the linkage to the book material become less clear. You might rightfully look at some of these examples and think to yourself, "Why didn't he just combine these two steps?" or "Isn't that step unnecessary?" The answer to both of these is likely, "This code illustrates the applications of the concepts the best."

Introduction to Quantitative Ecology: Mathematical and Statistical Modelling for Beginners. Timothy E. Essington,
Oxford University Press. © Timothy E. Essington 2021.
DOI: 10.1093/oso/9780192843470.003.0013

13.1 The basics

Most of your R coding work will be done in some form of a text editor. In a text editor, you'll write a series of commands following the syntax that R needs and then submit those commands to R to implement them. Rstudio provides a very nice integration of graphics, command window, and text files that most people find useful.

13.1.1 First, some orientation

Generally, when you use Rstudio, you will have multiple panels available. One panel is the *console*—take some time find to identify this panel now. The console contains the *command line*, or *prompt*, indicated by > and a blinking cursor (the cursor only blinks if you have activated this panel by clicking on it). If you type a command into the command line and hit Enter, R will run that command. So, if you type 2 + 3 into the command line and hit Enter, R will run this line of code and return 5. When you do this, the result will look like

```
> 2+3
[1] 5
```

Notice that [1] appears before the 5. This notation is just indicating the position of the first element of the answer. This is useful when the answer to the command spans multiple lines but is useless in the current example. In other words, the numbers in the brackets are not part of the answer; they are just there to help you read the answer. Later, more complex examples will make the benefit of this clearer.

There should also be other panels in Rstudio, including the environment panel that normally has tabs called "workspace" and "files," and then another panel that contains many tabs, including "help," "plots," and "packages." Once you create an R script (below) a fourth panel will appear as well.

13.1.2 Writing and running R code

Usually, you'll write your code in an "R Script" file. This is a text file that lists commands that you want R to do. In Rstudio, create a new script by going to the File menu, and selecting New /R Script. You'll see a blank panel appear (similar to what happens when you create a new module in Visual Basic). Type in the code below that creates two objects, called *a* and *b*, and assigns the value of 5 to *a* and 10 to *b*:

```
a <- 5
b <- 10
```

The first line does two things. It creates the object *a* and assigns the value of 5 to it, using the notation <- (this notation consists of two characters: the *less than* character and the *hyphen* character). Note that you can also use a = 5 to do the same thing. People have strong feelings about using the = notation. When you are beginning, use what is comfortable to you. The second line creates the object *b* and assigns the value of 10 to it.

Because this is in your script and so is not in R's memory yet, you need to run your code. You have several options available to you. The first, and probably least useful, is to literally copy the commands from your script, paste them into the console at the prompt, and hit Enter.

The second way is to run each line of your R script code directly. Place your cursor on the first line of your code, and then look at the Rstudio window for an icon that says "run." Click that, and you'll see a `<- 5` appear in your console. You can do the same thing to run the second line. You can also select and run multiple rows at a time (i.e., you can select both rows 1 and 2 and then select "run," and Rstudio will run those lines of code through the command window for you. This is a very convenient way to work with your R script.

The third way is to tell Rstudio to run the entire R script. On the R script window, there will be a button called "Source." If you select that, the entire set of commands in the R script will be run at once.

The fourth way is to save the file and type `source(filename)` from the command line. For instance, save your file as "test.R" and then, in the console, type `source("test.R")`. It will run the entire contents of the file, but you will not see each individual line of code repeated in the command window.

We can also ask R to return the value (or values) assigned to any object. Within the console, go to the prompt, type in a, and then hit Return, and R will return 5. You can also type in `print(a)` and then hit Enter, and it will do the same. You can also insert the commands `print(a)` and `print(b)` into your R script, so if you Source the file, it will assign the values to the objects and then show you the values that it assigned.

Finally, you can add text comments to your code. This is extremely useful for your own sake, and for others that might examine your code. You add comments using the # symbol, such that everything you type after the # within a line will not be interpreted by R. So, we might do something like this:

```
a <- seq(1,10) # this is a parameter
b<- seq(21,30) # this is another parameter
print(a) # this prints the output
```

This way, when reviewing your code, you have annotations telling you what each line is doing.

Of course, R has many many built-in functions, and it's impossible to list all potentially useful ones. In addition to those that are built into the base installation of R, there are many extra packages you can install, each of which has even more functions. Here I'll only explore those that come with the base installation of R, and ones that are particularly useful for modeling. We'll learn more functions in examples later on in the book.

`c()`
The `c()` function literally means "combine," so it is a way to string together several values into one vector. So, if we were to type `c(1,2,3,4)`, we would be combining values 1,2, 3, and 4 into a single object.

```
eq(from = , to= , by = )
eq(from = , to = , length.out = )
```

This generates a sequence starting at "from" and ending at "to." The input command "by" tells R the increment size. Alternatively, you can specify "length.out," which is how long you want the resulting sequence to be. You can't specify both "by" and "length.out."

Suppose you wanted to generate a vector of numbers that went from 1 to 5 in increments of 0.5:

```
> seq(from = 1, to = 5, by = 0.5)
[1] 1.0 1.5 2.0 2.5 3.0 3.5 4.0 4.5 5.0
```

Notice that `seq` can be used in place of `c()` in some cases. Above, the command `c(1,2,3,4)` could be replaced by `seq(from = 1, to = 4)`, as the default value of "by" is 1.

A special version of `seq` is to say X:Y. This will generate integers from X to Y. So the above could be shortened to 1:4.

You can refer to a single element of an array or vector using the `[]` command. For instance, if I wanted to refer to the first element of an array x, I would type the following:

```
>x <- seq(from = 1, to = 5, by = 0.5) # create array
> x[1] #print the first element in the array x
[1] 1
```

You can also refer to multiple elements of an array or vector at the same time:

```
>x <- seq(from = 1, to = 5, by = 0.5) # create array
> x[c(1,3,5)] # print out the 1st, 3rd, and 5th elements of
the array
[1] 1 2 3
> x[1:3] # print out the first three elements of the array
[1] 1.0 1.5 2.0
```

`length(x)`, `nrow(x)`, `ncol(x)`
These are all somewhat similar. For vectors, `length(x)` tells you how long the vector x is. For matrices, `nrow()` and `ncol()` will tell me the number of rows and columns, respectively. Below, I create an object called x and then assign to it a vector that is a

sequence from 1 to 5 in increments of 0.5. I then use length() to determine how long this vector is:

```
> x<-seq(from = 1, to = 5, by = 0.5)
> length(x)
[1] 9
```

I can refer to the last element of vector x by using

```
x[length(x)]
[1] 5
```

rep(x, times=n)
This takes the input argument x and repeats it n times. The input argument could be a vector or a single value or character. I often use this to create an empty vector that I will later populate to store model output. An empty matrix here means that I fill a matrix with NA, which stands for "Not Available":

```
> rep(x = NA, times = 10)
 [1] NA NA NA NA NA NA NA NA NA NA
```

which(logical statement)

This is a convenient way to find which elements of a vector meet some condition. Say we had a vector x that takes values from 11 to 20. We might want to identify which of the elements of x have values that are equal to or exceed 15. We would do it like this:

```
> x <- 11:20
> which(x>=15)
[1]   7   8   9 10 11
```

If you were to create an object using the command x>=15, you would get a vector that contains TRUE and FALSE for each element, corresponding to whether the condition x>15 was met.

13.1.3 Statistical functions

There are many different statistical functions. Basically, all of the probability distribution in this book are covered. Generally, to generate random numbers, you refer to the function that has an "r" in front of it. For instance, to generate ten random Poisson random variables, where the product of r times t equals 20, you would use

```
> rpois(n = 10, lambda = 20)
 [1] 25 24 20 14 22 17 24 21 10 29
```

Use `rbinom` for binomial random variables, `rnbinom` for negative binomial random variables, `runif` for uniform random variables, `rnorm` for normal random variables, and so on.

To get the probability or probability density given an observation, you'll use the function that has a 'd' in the front of it. To calculate the probability of each of the observations calculated above, use

```
>x <-  rpois( n = 10, lambda = 20)
>dpois(x, lambda = 20)
 [1] 0.044587649 0.055734561 0.088835317 0.038736640
 [5] 0.076913695 0.075954196 0.055734561 0.084605064
 [9] 0.005816307 0.012515304
```

Note that when you run these commands on your own computer, you will get answers that are different from those listed above because the vector x is random.

13.1.4   Basic plotting

R is capable producing outstanding scientific graphics. Entire books have been written on how to use R's impressive graphics capability (just search for "R graphics book" and you'll see many). Here we will review the basics, that is, the types of plots that are useful for displaying and evaluating model output. I will use the base plotting commands, but note that a package called `ggplot2` is very popular and provides different ways of creating graphics.

Replace your code above with the following:

```
a <- seq(from = 1,to = 10)
b <- seq(from = 21,to = 30)
```

This is assigning to a a sequence of integers that start at 1 and increase to 10, and assigning to b a sequence of integers that start at at 21 and increase to 30.

We want to plot a versus b. The code is fairly simple and intuitive:

```
plot(x = a, y = b)
```

This will create a plot where each a, b pair is a single point. We can ask R to make a line instead:

```
plot(x = a, y = b,
type = ''l'')
```

This says to use the type "l" which stands for "line." The default gives type "p," which is "point." If you want both, use `type = "b"`.

You can do much more, such as set x and y labels, set minimum and maximum axis values, set axis ticks, set the color and thickness of lines, change the symbol shape, and so on. Again, the internet is full of worked examples.

13.1.5 Data input and output

Often, we need to pull data into R or save R output into some other file type. Before we do that, we need to review the basics of working directories. If you ask R to retrieve a file, it will look in what is currently the "working directory" on your computer. You can retrieve the current working directory using the getwd() command. If you want to change the working directory, you can enter setwd("working directory name"), where you type in the full path to your working directory in quotations (e.g., "C:/users/timessington/quantecol_book/R for PC, /users/timessington/quantecol_book/R for Mac)".

Here we'll create a comma-delimited file in a spreadsheet, save it in a working directory, and then import this into R.

Open a spreadsheet. In cell A1, type in "a," and in cell B1 type in "b." These will be used as our column headings. Then, in cells A2 to A11, type in values $1, 2, 3, \ldots, 10$. In cells B2 to B11, type in values $21, 22, 23, \ldots, 30$. Now, go to File /Save as, and a window will appear. First, you'll need to find a place to put this file. You might find it useful to create a folder called "R_code" that is somewhere easy to find (perhaps on your desktop). Now you need to assign a name to this file. Call it something easy to remember, like "testdata." Finally, you need to tell your spreadsheet the file format you want to use for saving the file. Select "comma-delimited file" (CSV).

Now that you have a data file, let's see how we can input this into R.

Create a new R Script file. At the top of the file, use the setwd() command to set the working directory to the folder that has your data file in it. Then type in

```
thedata <- read.csv(file = "testdata.csv", header = TRUE)
```

This is creating an object called "thedata" and then using the function "read.csv" to read in the csv file. The function requires the filename (note I used "testdata.csv" as my filename, but this should match whatever you used to name your file). It also has the option to specify that the data file has a header row that contains the parameter names. So, by using header = TRUE, I'm saying to use the first row for header names.

Run this R script using whichever method you choose, and then, in the console, ask R to print out the object thedata so you can see the results. You should see an object returned that has ten rows and two columns, and the columns are named "a" and "b."

Note that there are many ways to input data into R; this is just one way.

You can export R objects as data files as well. Suppose we input the data file testdata.csv, assign that to the object thedata, and modify the object by creating a third column called column c which is the sum of columns a and b. We can do that in R using thedata$c <- thedata$a + thedata$b. Note the use of the $a operator which says, "Use the column named a." We can save this as a new .csv file named testdatawithc.csv, as follows:

```
write.csv(x = thedata, file = ''testdatawithc.csv'')
```

where `x = thedata` tells R what object should be saved, and `file = "testdatawithc.csv"` gives the filename. This file will be saved in the current working directory.

13.1.6 Looping

Looping plays a central role in ecological modeling. It is how we tell R to repeat iterative calculations, such as projecting forward a population from one time step to a future time step. It is also a convenient way to solve thorny programming issues when you can't find a single clever command to do exactly what you want.

Looping is just way to repeat some series of commands for a certain number of times. The basic syntax of a loop that runs 100 times is

```
for (i in 1:100) {
do some commands
}
```

The first line tells R "we are going to loop" and also tells the looping specifications. It creates a new object called `i` that takes different integer values as the loop proceeds. Those different integer values are included after the `in` statement. In this example, it is using a vector of values that starts at 1 and continues to some specified value, here 100; `i` will take a value of 1 during the first iteration of the loop, a value of 2 during the second iteration, and so on.

The second line right now is a placeholder, just to show that this is where all of the looping commands belong. In most cases, you'll have multiple lines of code here.

Finally, the commands are encompassed by curly brackets.

Here is a simple example: suppose you wanted R to count to one million. That is, it will take an object called `count` and iteratively add 1 to it, repeating that operation one million times. Your code might look like this:

```
count <- 0
for (i in 1:1e+6) {
    count <- count + 1
}
```

where `1e+6` is a shorthand for 10^6, or 1 million. Go ahead and run this code and confirm that at the end, count equals one million.

13.1.7   **Loops within loops**

You can nest one set of loops inside another set of loops. Suppose you have vectors of values for each of two parameters; call them "x.list" and "y.list." We may want cycle through all of them, calculate some quantity, and save to some output.

To start, we'll create a blank object to store the results. I usually use the notation output, but you will likely develop your own naming conventions:

```
output <- matrix(data = NA, nrow = length(x.list), ncol =
    length(y.list))
```

This makes a matrix that is full of NA' and has the same number of rows as the vector x.list, and the same number of columns as the vector y.list. I aim to replace all of these with values through looping.

Suppose you want to add x and y together, divide that by x, and then save the result in the output matrix. Your loop might look like this:

```
for (i in 1:length(x.list){
    for (j in 1:length(y.list){
        output[i,j] <- (x.list[i]+y.list[j]) / x.list[i]
    }
}
```

The notation x.list[i] means to take the *i*th element of the vector x.list, y.list[j] means take the *j*th element of the vector y.list, and the notation output[i,j] <- says "assign the *i*th row, *j*th column of output."

13.2   **Practicum: A logistic population model in R**

As we did in section 12.1, we'll work though the logistic population growth model to see how might implement a dynamic model in R.

First, create a new R script in Rstudio. Just as in spreadsheets, it is useful to define your parameters as named objects at the top of your R script. So, begin as follows:

```
r <- 0.2
K <- 100
N0 <- 10
tmax <- 20
years <- 0:tmax
```

Note here that, in addition to the parameters *r*, *K*, and initial population size, I am also specifying how many years to simulate (tmax) and then creating a vector that lists the years, starting at 0 and going to tmax.

Next, create an object called output to store the results:

```
output <- rep(x = NA, times =length(years))
```

Next, assign the initial population abundance to the first row of the vector output:

```
output[1] <- N0
```

The command output [1]   < - is saying, "Assign the first element of the object output to the value that appears on the right-hand side," which is, in this case, the initial population size.

Use a loop to apply the logistic recursive equation and project population size from year t to year $t + 1$:

```
n.loop <- length(years)
for (i in 2:n.loop) {
  n.t <- output[i-1]
  n.t.plus.1 <- n.t + n.t * r * (1 - n.t/K)
  output[i] <- n.t.plus.1
}
```

First, this code calculates how many times the loop should iterate: the length of the vector years. Then, within the loop, it finds the value of N_t from the vector output in the previous time step. Using that, it calculates N_{t+1} by using the logistic recursive equation. Finally, it assigns this value to the ith element of the output vector. Because these calculations are in a loop, it will iterate through all of the time steps, calculating population size for each time step.

Some students find it confusing that we place N_{t+1} in the ith element of the output. If you wanted your code to look more like the model equations, you could equivalently index the elements of output as

```
n.loop <- length(years) -1
for (i in 1:n.loop) {
  n.t <- output[i]
  n.t.plus.1 <- n.t + n.t * r * (1 - n.t/K)
  output[i+1] <- n.t.plus.1
}
```

The final step is to create a decent plot:

```
plot(x = years, y = output,
     type = ''b'',
     xlab = ''Year'',
     ylab = ''Population Size'')
```

Here I specified the x and y labels, using the arguments `xlab` and `ylab`, respectively. You could also specify the x and y axis ranges using something like

```
plot(
  x = years,
  y = output,
  type = "b",
  xlab = "Year",
  ylab = "Population Size"
  xlim = c(0, 20),
  ylim =  c(0, K)
)
```

where the x range is set from 0 to 20, and the y range is set from 0 to K.

13.3   Creating your own functions

Functions are a common shortcut tool in programming languages. By using functions, we can replace several lines of code that need to be run multiple times with a simple function call. Also, it is often helpful to take advantage of the fact that "what happens in functions stays in functions." Suppose that, in the midst of the function calculation, R needs to create and use some variable called "t," but you already have a variable called "t" in the workspace. You don't want R to overwrite your variable "t" while it is doing the function call. Fortunately, because of the "what happens in functions stays in functions" policy, that function variable "t" will disappear as soon as the function is done and will never enter the environment.

The basic syntax to create a function is

```
fun.name <- function(input arguments) {
do some stuff
return(output)
}
```

Let's go back to our trivial example in section 12.5, where we wanted to take two variables, x and y, add them together, and divide by x. This would be

```
sumanddivide <- function(x,y) {
output <- (x + y) / x
return(output)
}
```

The return statement is important because, otherwise, the function doesn't know what should be replaced. Functions can return anything you want: a scalar, a vector, a list, and so on.

Because this is such a simple function, with only one line of calculation, we write it more compactly as

```
sumanddivide <- function(x,y) (x+y)/x
```

thus obviating the need for the curly brackets and the return statement.

Because this is so simple, you probably wouldn't even bother writing a function for it. But you can write a function to handle more complex calculations. Above we wrote a loop to project a logistic population model. Suppose we needed to do that multiple times in our R script. It would be annoying to have to write that out each time. Instead, we can write a function that will take the model parameters and return the output vector:

```
logistic.fun <- function(r, K, N0, tmax) {
  years <- 0:tmax
  output <- rep(NA, length(years))
  output[1] <- N0
  n.loop <- length(years)
  for (i in 2:n.loop) {
    n.t <- output[i-1]
    n.t.plus.1 <- n.t + n.t * r * (1 - n.t/K)
    output[i] <- n.t.plus.1
  }
  return(output)
}
```

This script names the function `logistic.fun`, lists the parameters it needs as input, does the calculations, and specifies which object to return. To call the function, we would type

```
logistic.fun(r, K , N0, years)
```

We might assign this output to some object, like

```
logistic.output <- logistic.fun(r, K, N0, years)
```

This might be helpful if we were looping through different values of the model parameters. For example, back in section 12.4, we cycled through different values of *r* and saved the final population size for each value. We can do that easily in R:

```
r.list <- seq (0.1, 2.0, by = 0.1)
n.20.vs.r <- rep(NA, length (r.list))

for (i in 1:length(r.list)) {
  logistic.output <- logistic.fun(r.list[i], K, N0, tmax)
  n.20.vs.r[i] <- logistic.output[21]
}
```

Here I created an empty vector called n.20.vs.r to save the results, I cycled through values of *r* listed in the vector r.list, ran the logistic function to return a vector of population sizes, and then saved the final year result in vector n.20.vs.r.

Also, we can use the function call directly in another function call. For instance, we could plot the model like this:

```
plot(years, logistic.fun(r,K, N0, tmax),
     type = ''b'',
     xlab = ''Year'',
     ylab = ''Population Size'',
     xlim = c(0,20),
     ylim =  c(0, K)
)
```

The bottom line is that functions are a useful way to clean up your code to simplify the syntax and organize the information flow. *Always debug the code within your functions extensively before using them.*

14

Skills for Dynamic Models

This chapter will provide worked-through examples that were used in Part 1. In most cases, separate instructions are given for spreadsheet and R. However, when some activities are far easier to do in programming environment than in spreadsheets, I'll only show the R instructions.

Note that the previous two chapters already gave us a head start on modeling in spreadsheets and in R. Namely, they showed how to run a logistic population model, given a starting population size and model parameters. The same logic used there will hold for other models as well. Namely, we first define the parameters and parameter values, and the starting levels for each state variable. We set up our spreadsheets and R code to use those parameter values and then apply a recursive equation to model the state variables at discrete-time steps. In spreadsheets, we handle discrete-time steps as distinct rows. In R, we handle discrete-time steps as elements in a vector.

14.1   Skills for population models

14.1.1   Implementing structured population models

Structured population models have more detail than their unstructured counterparts, but the underlying mechanics of coding and running them are basically the same. Regardless of whether you are using spreadsheets or R, you'll generally start by listing all of the parameters of the model and assigning values to each. In spreadsheets, we also saw that is is useful to name the cells containing parameter values.

This example will work through the example shown in section 3.1.1, using the parameter values shown in table 14.1 and the starting values shown in table 14.2.

14.1.1.1   Age structure in spreadsheets

Set up your spreadsheet by listing the parameters and parameter values up top as usual. Name the parameters (but note that you will not be able to assign names like "F3" because that is already the reference to column F, row 3. Instead, you can name such cells "F_3"). In columns to the right of the parameters, also list the state variable names and their starting levels (and name the cells containing the starting values).

Introduction to Quantitative Ecology: Mathematical and Statistical Modelling for Beginners. Timothy E. Essington, Oxford University Press. © Timothy E. Essington 2021.
DOI: 10.1093/oso/9780192843470.003.0014

Table 14.1 *Parameter values for an age-structured model*

Parameter	Value	Parameter	Value
F_3	10	S_1	0.2
F_4	15	S_2	0.3
F_5	20	S_3	0.4
		S_4	0.5

Table 14.2 *Initial conditions for an age-structured model*

State variable	Starting value
N_1	20
N_2	15
N_3	10
N_4	5
N_5	1

We use the same basic logic to run a structured population model as we did with an unstructured population model. The only difference is that we have more state variables to keep track of, which we do by allocating one row for each state variable.

In cell A10, type in "Year," and, in cells B10 through F10, type in "N1," "N2," "N3," "N4," and "N5." In cell A11, enter "0", and then fill in cells A12 down to A61 so that they contain the numbers 1, 2, 3, ..., 50, respectively.

Row 11 needs to be populated with the starting values of each age class. In cell B11, type in a formula so that the cell will return the starting level for age 1. For instance, if the cell containing the starting level for N_1 is in cell D1, you might type =D1 into cell B11. If you named this cell "N_1," then you might type =N_1. Repeat that procedure to input the starting levels of age 2, age 3, age 4, and age 5 in row 11.

In row 12, we finally start to use the recursive equations. Following eqn. 3.1, we know that the number of age 1 individuals is based on the numbers of age 3, age 4, and age 5 individuals and their respective fecundities. Assuming you named the parameter cells, the formula for cell B12 is

=F_3 * D11 + F_4 * E11 + F_5 * F11

Once you hit Enter, the cell should return "195". If it does not, check the formula, the parameter values, and the starting values. Take a moment to confirm why this formula does what it should.

Table 14.3 *Spreadsheet layout for an age structured model*

	A	B	C	D	E	F
10	Year	N1	N2	N3	N4	N5
11	0	20	15	10	5	1
12	1	195	4	4.5	4	2.5

Once this cell is behaving you can move over to cell C12. Here you need to calculate the number of age 2 individuals. This is equal to the number of age 1 individuals in the previous time step, times the age 1 survivorship. Thus, in cell C12, your formula is = **B11 * S_1**. Similarly, you calculate the number of age 3 individuals into cell D12 as =**C11 * S_2**, the number of age 4 individuals into cell E12 as =**D11 * S_3**, and the number of age 5 individuals into cell F12 as =**E11 * S_4**.

If these have all been entered correctly, you should obtain something that looks like table 14.3.

Frequently, students will enter something like **N_1 * S_1** in cell C12, thinking that the number of age 2 individuals equals the initial number of age 1 individuals times their survivorship. This will give the right answer in cell C12 but will be completely wrong for all other time steps (rows). By using the relative cell reference to the population size in the previous row (instead of the named initial condition), you can make the spreadsheet repeat the calculation for each time step when you copy the formula down the rows.

Select cells C12 through F12, copy, and then paste into cells C13 through F61. You should now have the full model up and running, with final (year = 50) population sizes of 1287.95, 245.15, 69.99, 26.65, and 12.68 for age 1 though age 5 (note: I rounded these values so you may not see the exact same answer down to the 0.01 decimal place).

You can now calculate the population growth rate by examining the numbers of individuals in the population in the last and second-to-last time steps. Add together the number of individuals in the entire population in column G (e.g., in cell G11, you would type =**B11 + C11 + D11 + E11 + F11**). The population growth rate can be estimated as the ratio of the population size in year 50 divided by the population size in year 49. Assuming that column G contains the total population sizes for these years, this would be calculated as =**G61 /G60**.

You can also calculate the age distribution of the population, that is, the proportion of individuals within each age class. The proportion of individuals that are age 1 is calculated as = **B61 /G61** (this is the number of age 1 individuals divided by the total number of individuals in the population). Repeat this to obtain the population age structure for all age classes.

14.1.1.2 *Structured population modeling using matrices*

We saw that we can represent the series of equations of the age-structured model as a single equation using matrices. We can repeat our model calculations above by first generating the transition matrix **A** and then applying matrix algebra to calculate the vectors of population sizes for each year.

To create the transition matrix, we need to define a five-row by five-column area that contains the matrix shown in eq. (3.17). Suppose we will do this in cells E1 through I5. I find it easiest to begin by entering 0s into each cell, to begin the matrix, and then go into the cells that contain the nonzero values and change them as needed.

Start your transition matrix by entering 0s in cells E1 through I5. Then work through the nonzero elements of the matrix by referencing the parameter values. For instance, in cell G1, type =F_3; in cell H1, type =F_4; and, in cell I1, type =F_5. Complete the matrix by filling in the parts of the matrix that contain the survivorship parameters.

Above we learned that we can name cells. We can also name entire ranges of cells! This is helpful because we can define the cell range that contains the transition matrix as **A** and then refer to this name in our spreadsheet formula. Select cells E1:I5 and then type in **A** into the box where you name cells and hit Enter.

We do not need to change the first row of the model (the one containing the initial starting sizes). But we will change the formula for all rows that apply the model recursive equation (here beginning in row 12). We immediately face a small problem. The notation in equation 3.19 presumes that the population densities are a vector with five rows and one column. But, in our spreadsheet, the population densities are shown as a vector with one row and five columns. So, we will need to use the **transpose**() function in our spreadsheet to flip the vector of population sizes.

It is easier to understand by first entering in the spreadsheet formula and then follow the explanation. Select cells B12:F12, and then, in the formula bar, type in the array formula (remembering not to hit Enter to submit it!):

=TRANSPOSE(MMULT(A, TRANSPOSE(B11:F11)))

To submit the formula, use Ctl-Shift- Enter (see section 12.3). You will know if it worked if it returns the correct vector of population densities as you had before. If it is not working, that's alright. First let's work though what this formula is doing.

The core operation we are tying to do is to multiply the matrix **A** times the matrix N_t. The part of the formula that includes **=MMULT**() is doing just that. To get the vector for N_t, we take the range of numbers in cells B11 through F11 and transpose it so that it has five-rows and one-column. Thus, the inner part of the function (**MMULT(A, TRANSPOSE(B11:F11)))** is generating a five-row by one-column vector of N_t, multiplying that by **A** to create N_{t+1}. We want to place the result into a single row with five columns, so we need to take take the transpose of that operation. That is what the outer **TRANSPOSE**() function is doing.

Some people find the matrix version of the model easier because there is a single equation for each time step. Others find the quirks of array formula to be distracting and difficult. There is no right or better approach.

Calculating eigenvalues and eigenvectors in spreadsheets is not straightforward. It can be done, but, by the time you figure out how to do it and then implement it, you may as well have calculated it more easily in R or another program that can do it for you.

14.1.1.3   *Age structure in R*

Open a blank R script, and assign the parameter values from table 14.1 and the starting values from table 14.2. You can use any notation you wish, but I often denote starting values using ".start" after the state variable name. So I might use `N1.start <- 20` to assign the starting value of 20 to age 1. I find it useful to define `tmax` as the maximum time step, and then define a vector of years by `years <- 0:tmax`.

Next, we need to create a blank matrix to store the model output. We will run the model from time step 0 to time step 50, as we did with the logistic model previously. As before, we can call this matrix "output," which we can create as follows:

```
output <- matrix(NA, nrow = length(years), ncol = 5)
```

This creates a matrix of 51 rows and 5 columns, that is currently full of NA. The code will replace the NA with the model output.

We begin by assigning the starting population sizes in the first row of `output`:

```
output[1,] <- c(N1.start, N2.start, N3.start, N4.start,
    N5.start)
```

The notation on the left-hand side `output[1,] <-` means "assign to all columns of the first row of output." The right-hand side is just a vector of starting population levels.

We then use a loop to cycle through the population recursive equation. As we did for the unstructured population model, we loop as follows:

```
n.loop <- length(years)
for (i in 2:n.loop) {
    output[i, 1] <- F3 * output[i-1,3] + F4 * output[i-1,4] +
        F5 * output[i-1,5]
    output[i, 2] <- S1 * output[i-1,1]
    output[i, 3] <- S2 * output[i-1,2]
    output[i, 4] <- S3 * output[i-1,3]
    output[i, 5] <- S4 * output[i-1,4]
}
```

This code calculates the value of state variable in time "i" based on the values of the state variables in time "i − 1". For instance, the number of age 1 individuals in year 2 equals the number of age 3, age 3, and age 5 individuals in year 1, each multiplied by their respective fecundities.

To view the model output and check for errors, go to the console, and, at the prompt, type `View(output)`. This will open a window in the R script panel that shows the output matrix as you might view it in a spreadsheet. Check the second and final row numbers against those shown in section 14.1.1.1.

Calculate a vector of total population sizes by using the rowSums command ntotal <- rowSums(output) (rowSums adds up all of the values within a row of a matrix and returns the result for all rows).

To calculate the population growth rate, we take the number of individuals in the final time step and divide it by the number of individuals in the previous time step:

```
lambda <- ntotal[51] / ntotal[50]
```

or, more generally,

```
lambda <- ntotal[length(ntotal)] /
   ntotal[length(ntotal) - 1]
```

To calculate the age distribution, we simply divide the last row of the output matrix by the last row of the ntotal vector:

```
age.distribution <- output[nrow(output), ] / ntotal[length(ntotal)]
```

14.1.1.4 *Structured population modeling using matrices*

If we wish to replace the five equations with a single matrix operation and use properties of the transition matrix to calculate the population growth rate and stable age distribution, we must first create the transition matrix Equation 3.17.

We begin by creating a five-row, five-column matrix full zeros:

```
A <- matrix(0, nrow = 5, ncol = 5)
```

I like to do it this way because most of the elements are 0. So, it is easier to start with every element equal to 0 and then replace the ones with parameters as needed.

We then replace the nonzero elements of **A** with the model parameters to generate the matrix that appears in Equation 3.17. For instance, to place the F_3 parameter into the first row, third column, we type A[1,3] <- F3. Repeat this step to fill in the F_4 and F_5 parameters.

We then place the survivorship parameters into the matrix. Age 1 survivorship appears in the second row, first column: A[2,1] <- S1. Age 2 survivorship is in the third row, second column, age 3 survivorship is in the fourth row, third column, and so on. Thus, to create our transition matrix we use

```
A <- matrix(0, nrow = 5, ncol = 5)
A[1,3] <- F3
A[1,4] <- F4
A[1,5] <- F5
A[2,1] <- S1
A[3,2] <- S2
```

```
A[4,3] <- S3
A[5,4] <- S4
```

To view the transition matrix, go the console and, at the prompt, type A and then hit enter, and R will return the matrix for you to see:

```
> A
      [,1]  [,2]  [,3]  [,4]  [,5]
[1,]   0.0   0.0  10.0  15.0    20
[2,]   0.2   0.0   0.0   0.0     0
[3,]   0.0   0.3   0.0   0.0     0
[4,]   0.0   0.0   0.4   0.0     0
[5,]   0.0   0.0   0.0   0.5     0
```

Notice again the numbers in brackets. Their purpose is more obvious now that our answer is a matrix. At the top of the output, they are simply labeling the output so that you know which column each belongs to. For instance [,1] refers to the first column. At the left of the output, the numbers in the bracket are labeling the rows (e.g., [1,] refers to the first row).

To use matrix notation in your model, replace the model loop with the following:

```
n.loop <- length(years)
for (i in 2:n.loop){
  output[i,]<-  A %*% output[i-1,]
}
```

where the notation:

```
%*%
```

means to perform matrix multiplication.

R did something very useful for us here. We want to multiply **A**, which is a five-row, five-column matrix, by a vector of population densities that has five rows and one column, returning a new vector of population densities that has five rows and one column. But, above, our population densities have one row and five columns. Three things happened. First, output[i-1,] is a vector that has no real dimension. That is, R treats it as a vector of five numbers but doesn't assign it to be either a row vector (e.g., one row with five columns) or a column vector (five rows and one column). Second, R takes a vector (like the one created by output[i-1,]), and promotes them to be either a row vector or a column vector, whichever is needed to make it

continued

conformable with the matrix operation in hand. Third, R is happy to take the resulting product (here, literally a column vector) and turn it back into a vector without the "row" or "column" designation. In other words, it is figuring out a lot of the matrix orientation for us!

Now that you have created the transition matrix, you can quickly calculate the eigenvalues and find the largest one. In R, you do this as follows:

```
ev <- eigen(A)
eigen.vals <- ev$values
eigen.vects <- ev$vectors
```

The first line creates an object called `ev` and assigns the result of applying the `eigen` command to the transition matrix. The object `ev` contains both the eigenvalues and eigenvectors. The next two lines therefore extract out the eigenvalues or eigenvectors and assigns them to their own objects.

To see the eigenvalues, go to the console and type at the prompt `eigen.vals`. This will return a vector of all five of the eigenvalues. Notice that some elements of this vector are complex numbers (meaning you will see eigenvalues such as $-0.0768134 + 0.7400614i$). We only care about the largest one. Fortunately, R always provides the eigenvalue outputs in descending order so that the largest eigenvalue (and the corresponding eigenvector) always takes the first position. Therefore, we can get the population growth rate as

```
lambda <- Re(eigen.vals[1])
```

The command `Re()` is used here to remove reference to the complex part of the eigenvalues and eigenvectors (which we do safely because, in this particular case, the complex part will be 0 for the eigenvalues and eigenvectors that we care about).

View the value of `lambda` for these parameters: it should be nearly identical to the value you calculated above!

We can also extract the stable age distribution from the eigenvector. First, we need to extract the eigenvector that is paired with the largest eigenvalue:

```
w <- Re(eigen.vects[,1])
```

To get the proportion of individuals in each stage, we divide each element of vector w by the sum of the elements of vector w:

```
stable.age <- w / sum(w)
```

Once again, this should return the same values as you calculated above.

14.1.1.5 Advanced: Calculating elasticities in R

We know from section 3.4.1 that the "left" and "right" eigenvectors of the transition matrix can be used to calculate the elasticities—the proportional contribution of each element of **A** to the population growth rate.

Begin by calculating the right eigenvectors and eigenvalues from the matrix **A**:

```
ev <- eigen(A)
lambda <- Re(ev$values[1])
w <- Re(ev$vectors[,1])
```

Then calculate the left eigenvalues and eigenvectors. This is done by asking R to calculate eigenvalues and eigenvalues of the transpose of **A**:

```
evl <- eigen(t(A))
v <- Re(evl$vectors[,1])
```

Next, calculate the scalar product, which is simply

```
sp <- sum(v * w)
```

The numerator of equation 3.28 requires that we multiply v_i (where i is row) by w_j (where j is column) for all permutations of i and j. This can be calculated using a single command:

```
vi.wj <- v \%o\% w
```

where the operation returns the "outer product," which is exactly what we want.

Presuming that we calculated and defined `lambda` as described in the previous section, the elasticity matrix equals

```
e <-  A  / lambda * (vi.wj / sp)
```

This returns a five-row, five-column matrix containing 0 where A = 0, and returns the elasticity of each nonzero element of 0 otherwise.

14.1.2 Cobwebbing

Cobwebbing is a way to visualize how dynamic behavior is created by the underlying model functions. While it is certainly possible to create a spreadsheet to create these visualizations (and I do so in my classes), doing so is not straightforward, and spending time producing those graphics is only worth it if you are already pretty good at manipulating spreadsheets. For that reason, I only provide R guidance for cobwebbing: it is far simpler to do in a scripted language.

This section will explain code used to produce cobweb plots for a logistic model. Once this is mastered, you can simply swap out a different model function to apply it to other models.

As always, begin by defining your model parameters up top. You will need r, K, and starting population size N_0. These can be anything, but it might be useful to start with the values you used in section 13.2. Run the model for longer, say, fifty years.

First, use the same code as you created before to generate the vector called "output" that has the population levels in each time step.

Second, you need to create a plot that shows the logistic function as the relationship between N_t and N_{t+1}. To do this, create a vector of 100 population sizes from 0 to 1.5 $\star$ K, and assign this to object n.t:

```
nt <- seq( from = 0, to = 1.5 * K, length.out = 100)
```

From this, calculate N_{t+1}:

```
ntplus1 <- nt + nt * r * (1 - nt / K)
```

Finally, plot this curve as follows:

```
# plot logistic function
plot(x = nt, y = ntplus1,
        type = "l",
        lwd = 3,
        xlab = "Nt",
        ylab = "N t+1",
        ylim = c(0, 1.5 * K),
        xlim <- c(0, 1.5 * K)
        )
```

This is creating a line plot, with a fairly thick line (lwd=3 means to increase the default line thickness by a factor of 3), labels the x and y axes, and sets the limits for them.

Now we will add onto this plot. First, we want to add the replacement line. This is done using the abline() function:

```
abline(a = 0, b = 1,
        col = "gray",
        lwd = 3)
```

This commands plots a line with a y-intercept of 0 and a slope of 1, colors the line gray, and thickens the line for good visibility.

Now comes the more complex part—adding in the lines showing the model trajectory. I find it most useful to add all of the vertical lines first and then add the horizontal lines second. First, think of all of the vertical lines. They connect two points, $x = N_t$, $y = N_t$, and $x = N_t$, $y = N_{t+1}$. We can plot all of them by using a loop, as follows:

```
# plot vertical lines
n.loops <- length(years) -1
for (i in 1:n.loops){
    xs <- rep(x = output[i], times = 2)
    ys <- c(output[i], output[i+1])
    lines(xs, ys,
          lwd = 1
          )
}
```

This loop cycles through all of the time steps, generates a vector of two x values that both equal N_t, and a vector of two y values that equal N_t and N_{t+1}, respectively. It then draws a line connecting the two points together.

The last step is to add the horizontal lines of the cobweb. They connect lines at two points $x = N_t$, $y = N_{t+1}$ and $x = N_{t+1}$, $y = N_{t+1}$. The code below repeats this for each N_t, N_{t+1} pair:

```
# plot horizontal lines
n.loops <- length(years) - 1
  for (i in 1:n.loops) {
    xs <- c(output[i], output[i+1])
    ys <- rep (x = output[i+1], times = 2)
    lines(xs, ys,
          lwd = 1
          )
}
```

You should now have a fully functional cobweb plot. Change the value of r (make it larger) to see interesting cobweb shapes.

14.2 Skills for multivariable models

14.2.1 Calculating isoclines

The isocline of any state variable, X, is a line or a curve depicting the combination of state variable values that makes the X dynamic equation equal 0. Here we will walk through the algebraic steps need to calculate isoclines, using the continuous-time logistic competition model (equation 4.5) as our example:

$$\frac{dX}{dt} = f(X,Y) = r_x X \left(1 - \frac{X + \alpha Y}{K_x}\right)$$

$$\frac{dY}{dt} = g(X,Y) = r_y Y \left(1 - \frac{Y + \beta X}{K_y}\right)$$

14.2.1.1   The X isocline

We create an equation that shows the combination of X and Y that makes dX/dt equal to 0. We are planning on drawing isoclines with X on the x-axis and Y on the y-axis, so we will rearrange the equation so that Y appears alone on the left-hand side of the equation, and X appears on the right-hand side . The series of algebra steps might look like this:

$$r_x X \left(1 - \frac{X + \alpha Y}{K_x}\right) = 0$$

$$1 - \frac{X + \alpha Y}{K_x} = 0$$

$$X + \alpha Y = K_x$$

$$Y = \frac{K_x - X}{\alpha} \tag{14.1}$$

This last equation says, for any given level of X, what level of Y will make dX/dt equal to 0. In this case, it's simply a line with a y-intercept of K_x/α and a slope of $-1/\alpha$. If we wanted to plot it, we would just create a list of values of X and, for each, calculate the corresponding value of Y, using the right-hand side of the equation. Notice that the isocline doesn't depend at all on r_x, as it had no bearing on the sign (positive or negative) of dX/dt.

We can use the same procedure to create an expression for the Y isocline. We start with the expression for dY/dt, set it equal to 0, and rearrange:

$$Y = K_y - \beta X \tag{14.2}$$

This equation says that the Y isocline is a line with a y-intercept of K_y and a slope of $-\beta$.

Using isoclines to solve for equilibrium

You may remember at some point you learned how to solve for two variables if you had two expressions containing those variables. You may be one of the 99.5% of the population who forgot how to do this once you finished algebra. No problem, a little refresher will jog your memory.

continued

The equilibrium occurs when both dX/dt and dY/dt equal 0. Our two isoclines tell us when each individually equals 0. Thus, the values of X and Y that make both dynamic equations equal to 0 is the intersection.

Because we know they must intersect at these two points, they have the same value of X (call it $X^\star$) at the intersection point. We also know this intersection point has the same value of Y, called $Y^\star$. So, we can simply set the isolines equal to each other:

$$\frac{K_x - X^*}{\alpha} = K_y - \beta X^* \tag{14.3}$$

We then rearrange to get $X^\star$ on the left-hand side, and all of the model parameters on the right-hand side:

$$X^* = \frac{\alpha K_y - K_x}{\alpha\beta - 1} \tag{14.4}$$

We then substitute this expression into either of the isoclines to get $Y^\star$:

$$Y^* = \frac{\beta K_x - K_y}{\alpha\beta - 1} \tag{14.5}$$

14.2.1.2 *The case of the disappearing state variable*

Suppose our equation for dX/dt was something like this:

$$\frac{dX}{dt} = aX - bX - cXY \tag{14.6}$$

You do the usual trick to get the X isocline:

$$cXY = aX - bX$$

$$Y = \frac{aX - bX}{cX} \tag{14.7}$$

But note here that the X's all cancel out, leaving us with

$$Y = \frac{a - b}{c} \tag{14.8}$$

The X isoclines are supposed to tell us the combination of values of X and Y that make dX/dt equal to 0, but there is no X in the equation. This means that the level of the state variable X has no bearing on whether dX/dt is positive, negative, or equal to 0.

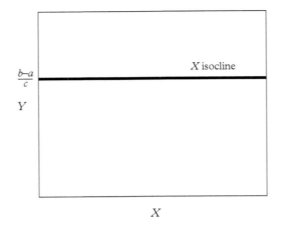

Figure 14.1 *When the state variable disappears from the equation, the isocline will be either a horizontal line (shown) or a vertical line.*

Instead, it depends only on Y. In this case, we would draw the X isocline as a horizontal line that intercepts the Y axis at the point $\frac{(b-a)}{c}$ (figure 14.1). Levels of Y above this line would make dX/dt negative (because of the negative sign on the term $-cXY$, we know Y is bad for X); levels of Y below this line will make dX/dt positive.

Finally, we also know that the equilibrium value of Y must be the value calculated in equation 14.8. This is seems counterintuitive, because the equilibrium value of Y is being determined from the expression for dX/dt. You can think of it this way: $Y^\star$ is the perfect level of Y that makes dX/dt unchanging. So, if Y were a predator on X, it is the abundance of predators needed to keep the prey, X, perfectly in check. We still don't know what $X^\star$ is, as that will be based on the Y isocline. The key point is that the equilibrium value might come from either isocline.

14.2.2   Calculating Jacobian matrices

Jacobian matrices are simply collections of derivatives of the dynamic equation functions with respect to the state variables. The eigenvalues of these matrices tell us a lot about the model behavior. You'll need to some a bit of calculus to do these, but a handful of derivative rules will usually do the trick.

Here we will calculate Jacobian matrices for the two-species competition model (equation 4.1). We need to calculate four derivatives, two for each state variable.

We begin with the dynamic equation for species X:

$$\frac{dX}{dt} = f(X,Y) = r_x X \left(1 - \frac{X + \alpha Y}{K_x} \right)$$

We first take the derivative of this expression with respect to state variable X, $\frac{\delta f(X,Y)}{\delta X}$.

It is often easiest to expand the expression so that the model function looks like a polynomial, where you are simply adding several terms together. Multiplying the term in parentheses by $r_x X$ gives

$$f(X,Y) = r_x X - \frac{r_x X^2}{K_x} - \frac{r_x \alpha Y X}{K_x} \tag{14.9}$$

Written this way, we can simply calculate the derivatives of each polynomial term with respect to X.

The first term's derivative is fairly simple. The derivative of the product $r_x X$ with respect to X is just r_x. If this isn't clear, then remember your derivative rules: the derivative of $a X^b$ equals $ab X_{b-1}$. Here $a = r_x$, and $b = 1$.

The second term is made easier if you consider r_x/K_x to be a single term, like a described above. The derivative of aX^2 equals $2aX$, and, by swapping in the full expression for a, you get $2r_x X/K_x$. The third term's derivative is calculated much the same way, leading to $r_x \alpha Y/K_x$. Many times I hear, "Wait, how come you are treating Y as a constant here when it is a state variable?" We can do it here because, at least for right now, we are only calculating the derivative of the function with respect to X. That is what the funny δ notation means. Soon, we will calculate the derivative of the function with respect to Y, where we will treat X as a constant.

Our final derivative is then

$$\frac{\delta f(X,Y)}{\delta X} = r_x - \frac{2r_x X}{K_x} - \frac{r_x \alpha Y}{K_x} \tag{14.10}$$

To calculate the value of the derivative, we need to apply the values for each of the parameters and specify values of X and Y to use in the calculation. Remember that we are interested in the model behavior near the equilibrium. So, we always use the equilibrium values of state variables when calculating the derivatives

For that reason, we sometimes seek to simplify the expression by substituting in expressions for X^* and Y^*. You can feel free to do that here, but nothing magical will happen; you will just have an expression for equilibrium that includes only the model parameters:

$$\frac{\delta f(X,Y)}{\delta X} = \frac{r_x}{K_x} \left(K_x - 2X^* - \alpha Y^* \right) \tag{14.11}$$

$$= \frac{r_x}{K_x} \left(K_x - \frac{2(\alpha K_y - K_x) + \alpha(\beta K_x - K_y)}{\alpha \beta - 1} \right) \tag{14.12}$$

You could do more math, but you likely won't land on on a vastly simplified expression. However, in some models, doing this quick algebra will lead to a very simplified expression (e.g., the Lotka-Volterra predator-prey model).

Now take the derivative of the X dynamic equation ($f(X,Y)$) with respect to Y. Again, take derivatives term by term, starting with equation 14.9. Right away, you will notice that

Y only appears in one term. What happens when Y doesn't appear? Remember that the derivative of aY^b equals abY^{b-1}, and when Y doesn't appear in a term, it is equivalent to having Y^0 in that term (i.e., $b = 0$). Applying the derivative rule, you end up multiplying the term by 0. So, we can simply ignore all of the terms that do not have Y in them because their derivatives will equal 0. Our derivative at equilibrium then becomes

$$\frac{\delta f(X,Y)}{\delta Y} = -\frac{r_x \alpha X^*}{K_x} \tag{14.13}$$

$$= -\frac{r_x \alpha (\alpha K_y - K_x)}{K_x(\alpha\beta - 1)} \tag{14.14}$$

We repeat the same procedure to calculate the derivatives of the Y dynamic equation with respect to both X and Y. They are

$$\frac{\delta g(X,Y)}{\delta X} = -\frac{r_y \beta Y^*}{K_y} \tag{14.15}$$

$$\frac{\delta g(X,Y)}{\delta Y} = \frac{r_y}{K_y}\left(K_y - 2Y^* - \beta X^*\right) \tag{14.16}$$

Note here I left X^* and Y^* in the expressions.

14.2.2.1 *What about discrete-time models?*

Suppose our model was in discrete time, expressed as the recursive equations

$$X_{t+1} = f(X_t, Y_t) = X_t + r_x X_t\left(1 - \frac{X_t + \alpha Y_t}{K_x}\right) \tag{14.17}$$

$$Y_{t+1} = g(X_t, Y_t) = Y_t + r_y Y_t\left(1 - \frac{Y_t + \beta X_t}{K_y}\right) \tag{14.18}$$

We would calculate the derivatives in the same way. In fact, there is really only one additional term present: the X_t and the Y_t on the right-hand side of the equations. Because the derivative of X_t with respect to X_t equals 1, this simply means that the derivative at equilibrium equals

$$\frac{\delta f(X,Y)}{\delta X} = 1 + \frac{r_x}{K_x}\left(K_x - 2X^* - \alpha Y^*\right) \tag{14.19}$$

Note that the derivative $\frac{\delta f(X,Y)}{\delta Y}$ will be unchanged. The same is true for $\frac{\delta g(X,Y)}{\delta X}$. The only other derivative that changes is $\frac{\delta g(X,Y)}{\delta Y}$, which will also have +1 added to it to account for the derivative of Y_t.

14.2.2.2 Calculating eigenvalues

You calculate eigenvalues in the same way as was done in section 14.1.1. The only differences are that we need to look more closely to determine which of the eigenvalues is largest and that we are interested in the real and complex parts of the eigenvalues.

Suppose in R you've created your Jacobian matrix and called it **A**. You would retrieve the eigenvalues using

```
ev <- eigen(A)$values
```

For continuous-time models, stability is determined only by the eigenvalue that has the largest real part:

```
max.eigen <- max(Re(ev))
```

For discrete-time models, you need to calculate the eigenvalue magnitude $\sqrt{r^2 + c^2}$, where r is the real part of the eigenvalue, and c is the imaginary part:

```
mags <- sqrt(Re(ev)^2 + Im(ev)^2)
max.eigen <-   max(mags)
```

where `mags` is a vector of eigenvalue magnitudes.

14.3 Monte Carlo methods

Monte Carlo methods are computer-based approaches to solving difficult mathematical or statistical problems. In essence, all Monte Carlo methods involve some sort of random sampling, performed many times, to calculate distributions of model outputs. The phrase "Monte Carlo" pays tribute to the glamorous gambling destination and the "luck of the draw" aspect of this method.

14.3.1   Monte Carlo example: What is π?

We know that the area of a circle with radius r is πr^2, while the area of an $r \times r$ square is r^2. The ratio of these two is therefore π. Suppose we wanted to use this relationship to get an estimate for π. One way to do that is through a Monte Carlo simulation. Essentially, we will virtually create a circle and square with radius = 1 and side = 1, respectively, both centered at $x = 0$ and $y = 0$. We will then take many random draws of x and y within -1 and 1. For each random draw, we calculate whether the x, y coordinates are within the bounds of the circle and the square. We will add up the number of times the random draw was within the circle, and the number of times the random draws was within the square. The ratio of these two should converge on π if we repeat this many times.

Here is some simple R code to perform this Monte Carlo simulation:

```
in.circle <- function(x,y) {
  return(ifelse(sqrt(x^2+y^2)<= 1, 1, 0))
}

in.square <-  function(x,y) {
  ifelse(abs(x) <= 0.5 & abs(y) <= 0.5, 1, 0)
}

n.iters <- 1000000
output <- matrix(NA, nrow = n.iters, ncol = 2)

for (i in 1:n.iters) {
  x <- runif(n = 1, min = -1, max =1)
  y <- runif(n = 1, min = -1, max =1)
  output[i,1] <-in.circle(x,y)
  output[i,2] <- in.square(x,y)
}
sum(output[,1]) / sum(output[,2])
```

The top two functions take the x and y draw and ask whether it is within the circle (in.circle) or within the square (in.square). It then runs 1,000,000 iterations, each time drawing a random value for x and y between -1 and 1, passing the random draws to each function and saving the resulting outputs (as either 0s or 1s) in the matrix output. Finally, it adds together all of the times the random selection was inside the circle and divides that by the number of times the random selection was inside the square.

The result? We get 3.142188, which is very close to the true value of 3.141593.

We use Monte Carlo methods in stochastic population models (chapter 5) because the population growth rate is random. We use Monte Carlo analyses for sensitivity analysis (chapter 15) because we want to explore the model output in high-dimensional parameter space. We use Monte Carlo analysis to propogate uncertainty in model parameter to express uncertainty in model outputs (chapter 5). And we use Monte Carlo methods in Bayesian methods to generate posterior probability densities (section 10.5).

This section will demonstrate the application of Monte Carlo methods in general. We will presume, for the sake of simplicity, that the goal of the Monte Carlo simulation is to repeat the calculation of a simple model under a several different alternative parameter values (this example will be given more context in chapter 15).

14.3.2 Monte Carlo simulation of population models

Harbor porpoises have naturally low population growth rates, and they are incidentally captured in fishing gear (van Beest et al. 2017). There is concern about the future prospects of these populations, so you have been asked to use the available information to assess risks and prospects for recovery. Here you will use a simple discrete-time density-independent model (equation 2.8), and then use a Monte Carlo analysis to asses the ranges of outcomes that are plausible.

We modify the basic density-independent model to separate deaths due to natural causes, which occur at the rate dN_t, and deaths that occur due to entanglement with fishing gear, fN_t. The dynamic model is then

$$\frac{\Delta N}{\Delta t} = (b - d - f)N \qquad (14.20)$$

where b is per capita birth rate per time step, d is the per capita death rate from natural causes, and f is the per capita death rate from fishing. Recall that this can be turned into a recursive equation:

$$N_{t+1} = (1 + b - d - f)N(t) = \lambda N(t) \qquad (14.21)$$

Under this density-independent model, we can project population size any n time steps into the future with

$$N_{t+n} = N_t \lambda^n \qquad (14.22)$$

Essentially, what we are going to do is treat the variables like outcomes in a slot machine: each time we run the model, we're going to get a different set of parameter values based on some random selection. This means that we need to come up with some rules for the slot machine ... how will the parameter values be drawn? In statistical jargon, we need to specify the probability function (chapter 7).

Table 14.4 lists plausible ranges of values for each parameter, based loosely on van Beest et al. (2017).

Table 14.4 *Plausible ranges for each model parameter*

Parameter	Lower	Upper
b	0.25	0.45
d	0.08	0.15
f	0.01	0.05
$N(1)$	75	125

Suppose we are interested in calculating the percent change in population size from now to twenty-five years from now. So, the metric of interest is

$$\text{percent change} = \frac{N(t=1) - N(t=25)}{N(t=1)} \times 100 \tag{14.23}$$

For now, we will assume that the parameters can take any number between the minimum and maximum value (i.e., we will assume a uniform probability density function). The steps that you would take are generally identical if you are assuming some other probability function; only the command for the random draws are different.

14.3.3   Spreadsheet guidance

First, you need to have a spreadsheet model that takes a single value for each parameter and calculates the population in twenty-five years; you can set it up as follows

1. First, define your model parameters up top as usual and put in placeholder values for each (it's good to choose values in the middle of the range). Don't forget to name the cells containing your model parameters (figure 12.1)!
2. Calculate λ.
3. Using this λ, calculate population size in $N_{t=25}$. You may use either the recursive equation (equation 14.21) or equation 14.22.
4. Calculate the percent change using eq. (14.23). Give the name "Pop_change" to the cell that contains this value.

Only when you are sure that your model is working (Is it? Have you debugged it?) should you proceed to make the parameters random draws. Debugging a model when the parameters are constantly changing is nearly impossible.

Below is a generalized algorithm to generate a uniform random number between a lower bound and an upper bound:

$$= (\text{upper} - \text{lower}) \times \text{RAND}() + \text{lower} \tag{14.24}$$

where RAND() is a uniform random number between 0 and 1. Take a moment to see why this makes sense: at the extreme value of RAND() = 0, eq. (14.24) will equal the lower bound, and, at the extreme value of RAND() =1, eq. (14.24) will equal the upper bound.

Place these lower and upper bounds on your spreadsheet, next to their accompanying named parameter value cells. Finally, replace the placeholder value in each with a spreadsheet equation that will apply equation 14.24. For instance, if cell B1 contains the value that is named '*b*', and cells D1 and E1 contain the minimum and maximum values, you enter in cell B1 the formula: **= (E1 - D1) * RAND() + D1**. You can then copy that formula down into all of the cells that contain parameter values.

Congratulations, now you have a model that generates random parameter values! Now you only need to automate things a bit. If you had loads of free time, you could manually recalculate the model, record the answer, and repeat a thousand times. But we know that a little Visual Basic code can do this for us.

Before doing anything, first make sure that your spreadsheet contains both Sheet1 (which is where your model calculations currently exist) and also a Sheet2, which is where you will store the Monte Carlo output. Once confirmed, open up the Visual Basic editor (section 12.5), go to the "insert" menu, and then select "module." On the first line, type

```
Sub Monte_Carlo ()
```

Visual Basic will automatically insert the End Sub line at the bottom. Your commands will be typed-between these two lines. We need to create a loop that will generate parameter values 1,000 times and save the output. Our code will look something like this if we named our cells b, d, f, N_1, and Pop_Change:

```
For i = 1 To 1000
    Sheet1.Calculate
    b = Range("b").Value
    d = Range("d").Value
    q = Range("q").Value
    N_1 = Range("N_1").Value
    Population_Change = Range("Pop_Change").Value

    Sheet2.Cells(1 + i, 1) = b
    Sheet2.Cells(1 + i, 2) = d
    Sheet2.Cells(1 + i, 3) = q
    Sheet2.Cells(1 + i, 4) = N_1
    Sheet2.Cells(1 + i, 5) = Population_Change

Next i
```

The first command of the loop tells Excel to recalculate everything on Sheet1. That includes drawing new parameter values and calculating the population trajectory. The next four lines simply record the values used for each parameter and look up the resulting population change.

The following five lines take these values and place them on Sheet2 of the workbook, where the values of b are listed in the first column, d in the second column, q in the third column, N_1 in the fourth column, and the model output in the fifth column. The first iteration will be returned in the second row, the second iteration in the third row, and so on. The last line tells Visual Basic when the looping commands end.

Lastly, you may want to put column headings on the model output on Sheet2. If so, you would add the following to the subroutine above:

```
Sheet2.Cells(1,1)="b"
Sheet2.Cells(1,2)="d"
Sheet2.Cells(1,3)="f"
Sheet2.Cells(1,4)="N(1)"
Sheet2.Cells(1,5)="Percent Change"
```

Run this model (either through Visual Basic or by running the macro via Excel), and view Sheet2 to examine the output.

14.3.4 R guidance

Your first steps are the same as they are for spreadsheet users:

1. First, define your model parameters up top and put in placeholder values for each (it's to choose values in the middle of the range).
2. Given the parameters b, d, and f, calculate λ.
3. Using this λ, calculate population size in N_{t+25}. You may use either the recursive equation (equation 14.21) or equation 14.22.
4. Calculate the percent change, using equation 14.23.

Only after you are sure this model is working should you move on to make the parameters random! parameters random!

First, create objects that define the range for each parameter:

```
brange <- c(0.25, 0.45)
drange <- c(0.08, 0.15)
rrange <- c(0.01, 0.05)
n1range <- c(75, 125)
```

You will also have to initialize a few other things:

```
# specify the number of monte carlo simulations
# and the number of years to simulate
nsims <- 1000
t <- 25
# setup matrix to store model output
output <- matrix(data = NA, nrow = nsims, ncol = 5)
colnames(output) <- c("b",
                      "d",
                      "q",
                      "N1",
                      "Percent Change")
```

Now you will run a simple loop that draws random values for each parameter, calculates the population size twenty-five years from now, calculates the percent change, and then saves it in the output matrix:

```
for (i in 1:nsims) {
  # draw parameter values from ranges
  b <- runif(n = 1, min =brange[1], max = brange[2])
  d <- runif(n = 1, min = drange[1], max = drange[2])
  f <- runif(n = 1, min = frange[1], max frange[2])
  n1 <- runif(n = 1, min = n1range[1], max = n1range[2])
  # Calculate lambda
  lambda <- 1 + b - d - f
  # calculate N 25 years from now
  n <- n1 * lambda ^ 25
  # calculate percent change
  nchange <- 100 * (n - n1) / n1
  # assign result to output matrix
  output[i, ] <- c(b, d, q, no, nchange)
}
```

14.4   Skills for stochastic models

This section will build skills in stochastic population modeling (chapter 5) through a series of steps. You will recreate the extinction risk calculations for Southern Resident killer whales (section 5.5). First, you will create a stochastic density-independent model, using the mean population growth rate, μ, equal to 0.0003 yr^{-1} and the standard deviation equal to 0.038. Then, you will run the model a thousand times and calculate the number of times that the population dropped below the quasi-extinction threshold. You will then modify the model to include autocorrelation to the r_t and uncertainty in the model parameters.

14.4.1   Stochastic models in spreadsheets

As usual, use the top left portion of your spreadsheet to define the parameters. You'll need to specify "mu" and "sd" (the mean and the standard deviation of the r_t's, respectively) and specify their values. Also, list the initial population size (here called N_0), assign the value of 73, and specify the extinction threshold value $N_{extinct} = 40$. As always, it is a good idea to name the cells. You should have parameter names in cells A1 through A4, and their accompanying values in cells B1 through B4.

In cell A9, type in "Year"; in cell B9, type in "rt"; and, in cell C9, type in "Nt." Fill cells A10 through A110 with 0, 1, 2, ..., 100.

In column B, we need to write a formula that will retrieve a random draw from a normal probability distribution. Use the procedure demonstrated in section 12.2, specifically, the =NORM.INV() command. In cell B10, enter the formula

=NORM.INV(RAND(), mu, sd).

When you hit enter, you should see a random number appear in the cell. Copy this formula down the B column, to cell B60.

Finally, in column C, we use a recursive equation to calculate population density at each time step. As usual, the population density in the first time step is the specified starting level. Presuming you named the cell containing this value N_0, the formula for cell C10 is **=N_0**. In cell C11, use the formula **=EXP(B10) * C10**. The first term is calculating λ_t, and the second term is multiplying that by N_t. Copy this formula down the C column down to cell C110. You now have a spreadsheet that runs a single iteration of a stochastic density-independent population model.

One last thing might be useful. When calculating extinction risk, we usually seek to determine the probability that the population drops below the quasi-extinction threshold. For that reason, it is convenient to create a formula that will return the minimum

Table 14.5 *Southern Resident killer whale counts by year*

Year	Count	Year	Count	Year	Count
1975	72	1990	88	2005	88
1976	71	1991	92	2006	89
1977	80	1992	91	2007	86
1978	80	1993	97	2008	85
1979	82	1994	96	2009	85
1980	84	1995	98	2010	86
1981	82	1996	97	2011	87
1982	79	1997	92	2012	84
1983	76	1998	89	2013	82
1984	74	1999	85	2014	78
1985	77	2000	82	2015	81
1986	81	2001	78	2016	83
1987	84	2002	79	2017	77
1988	85	2003	82	2018	75
1989	85	2004	83	2019	73

population size over the 100 years. That way, we can compare that to the quasi-extinction threshold to evaluate extinction risk. In cell A6, type in "Minimum Nt"; in cell B6, enter the formula **=MIN(C10:C110)**. Next, you will create a cell formula that returns a 1 if the model run resulted in a population size below the extinction threshold, and a 0 otherwise. In cell A7, type in "Extinct?", and in cell B7 enter **=IF(B6<Nextinct, 1, 0)**. This formula says to look up the value in cell B6, which lists the minimum population size in the simulation, and if that value is less than Nextinct, return the value 1; otherwise, return the value 0.

14.4.1.1 *Calculating extinction risk*

You can use your model above, along with a tiny bit of Visual Basic code to calculate extinction risk.

Your spreadsheet is set up to run one iteration of a population viability analysis. It runs the model once and returns 1 if the population drops below the extinction threshold, and 0 otherwise. The only thing left to do is to recalculate the spreadsheet a thousand times and, for each, return either 1 or 0. We learned how to automate such calculations in section 12.5. Follow those instructions to create a new Visual Basic module; at the top, type in sub PVA() and then hit Enter, and End Sub will automatically appear.

We want to recalculate the spreadsheet 1,000 (or more) times, and, for each, save the result. This means we write a loop that starts by instructing Excel to recalculate all of the values on this sheet, looking up the value that appears in cell B7 and creating a running count of the number of model simulations in which the population dropped below the quasi-extinction threshold. Our loop might look like this:

```
Sub PVA()
extinctCount = 0
For i = 1 To 1000
Sheet1.Calculate
extinctValue = Range("B7").Value
extinctCount = extinctCount + extinctValue
Next i
Sheet1.Cells(9, 7) = extinctCount
End Sub
```

This is simply recalculating the spreadsheet, looking up the value that is in cell B7 and assigning that to the object "extinctValue," adding that to the running count called "extinctCount," and then placing the final count in row 9, column 7 (cell G9). Note that this code presumes that all of your model calculations appear in the first sheet in your spreadsheet.

Run this macro any way you prefer. When you are finished, you should see a number appear in cell G9 that is the number (out of 1,000) of times that the population crossed the threshold.

You likely got a very small answer—under these parameters, the population rarely drops below the quasi-extinction threshold.

14.4.2   Stochastic models in R

As usual, create a blank R script document, and, at the top of the script, define and assign parameter values:

```
mu <- 0.0003
sigma <- 0.038
N0 <- 73
tmax <- 100
Nextinct <- 40
years <- 0:tmax
```

This assigns the population parameters, sets the initial population size to 73, specifies that you will run the model for fifty time steps, and assigns the value of 40 to an object called Nextinct—this is the quasi-extinction threshold.

It is useful to create the usual vector called "years" (helpful for plotting), as we have done before: years <- 0 : tmax. We then create vectors to store the N_t:

```
Nt <- rep(x = NA, times= length(years))
```

We generate a vector or r_t using the rnorm() function:

```
rt <- rnorm(n = length(years), mean = mu, sd = sigma)
```

As usual, we assign the starting value of population density in the first element of vector Nt:

```
Nt[1] <- N0
```

And then we create a loop that will calculate N_t by using the density-independent stochastic stochastic equation:

```
n.loops <- length(years)
for (i in 2:n.loops) Nt[i] <- exp(rt[i-1]) * Nt[i-1]
```

When calculating extinction risk, we usually seek to determine the likelihood that the population drops below a quasi-extinction threshold. Create an object that contains the minimum population size:

```
minpop <- min(Nt)
```

This way, you can readily compare the value of minpop to the quasi-extinction threshold:

```
extinct <- ifelse(minpop <Nextinct, yes = 1, no = 0)
```

This assigns the value 1 to the object extinct if the minimum population size is less than the quasi-extinction threshold, and the value 0 otherwise.

14.4.2.1 *Calculating extinction risk*

Your model above runs a single model iteration—it takes the model parameters, runs the model for fifty years, and returns the minimum population size. We want to repeat that many (1,000) times.

First, we need to create a vector output that will store the output for each model run. The only part of the model run that we want to store is whether the population dropped below the quasi-extinction threshold:

```
output <- rep( x = NA, times = 1000)
```

We now need to loop 1,000 times, each time calculating the object extinct, and saving its value to the output vector:

```
n.loops <- length(years)
for (j in 1:1000) {
  Nt <- rep(x = NA, times= tmax)
  rt <- rnorm(n = length(years), mean = rbar, sd = sigma)
  Nt[1] <- Nstart
  for (i in 2:n.loops) Nt[i] <- exp(rt[i-1]) * Nt[i-1]
  minpop <- min(Nt)
  extinct <- ifelse(minpop <Nextinct, yes = 1, no = 0)
  output[j] <- extinct
}
```

This code has two loops in it. The outer loop, indexed by j, assigns a blank vector of population sizes, generates a random draw of r_t's, and runs the inner loop. The inner loop, indexed by i, runs the population model. Once the inner loop is run, the code calculates the minimum population size and assigns 1 to the *j*th element of object "output" if it is below the quasi-extinction threshold, and 0 otherwise.

To calculate the number of model iterations in which the population dropped below the extinction threshold, sum the numbers in output (i.e., sum(output)). From there, simply divide this number by 1,000 to calculate the extinction probability.

14.4.3   **Advanced: Adding autocorrelation**

In section 5.5.1, you learned a formula (equation 5.15) to make the r_t autocorrelated. For a reminder, here it is:

$$r_t = \mu + \eta_t$$

$$\eta_t = \begin{cases} \rho\eta_{t-1} + \sqrt{1-\rho^2}v_t, & \text{if } t > 0 \\ v_t & \text{otherwise} \end{cases}$$

$$v_t \sim N(0,\sigma^2) \tag{14.25}$$

14.4.3.1 Spreadsheet guidance

First, add the parameter ρ to your spreadsheet. Type in "rho" in cell A8, and enter the value 0.23 in cell B8. Give the name "rho" to cell B8.

Second, add new columns to keep track of the v_t and the η_t. In cell D9 type in "eta" and type "vt" in cell E9. In cell E10, generate a random draw from a normal distribution with mean equal to 0 and standard deviation equal to 0.038:

=NORM.INV(RAND(), 0, sd)

Copy this formula down all the way to cell E110.

Now you need to calculate the η_t's. The calculation in the first time step is different from the others because we have to account for the fact that we don't have an η_{t-1} available to us. In cell D10, enter the formula **=E10**. In other words, for the first time step, we set η_t equal to whatever the random draw was in cell E10. In cell D11, you calculate η_t from the previous η_t and the current v_t to match equation 5.15:

= rho * D10 + SQRT(1 - rho^2) * E11

This is calculating each year's η_t from the previous η_t, plus the v_t multiplied by the correction term that ensures our overall variance is correct. Copy this formula down column D to cell D60.

Finally, we update the formula for column B to generate the r_t. In cell B10, replace the existing formula with

=mu + D10

This makes the current year r_t equal to μ plus η_t. Copy this formula down column B to cell B110.

You're all done. Just rerun your macro, and you'll get a new extinction risk estimate.

14.4.3.2 R guidance

First, add the parameter ρ to your R script, and assign the value estimated from the killer whale time series: `rho <- 0.23`.

Second, we will add new objects to keep track of v_t and η_t, and then update the code for the r_t vector. The η_t and r_t will be initiated as a blank vector (filled with NA), which we then populate as we calculate the model. When you want to assign the same thing to two or more objects, you can do it in a single line of code:

```
Nt <- rt<- etat <- rep(x = NA, times= length(years))
```

which assigns a blank vector of 60 NAs to the objects `Nt`, `rt`, and `etat`.
    Create the vector `vt` as

```
vt <- rnorm(n = length(years), mean = rbar, sd = sd)
```

Initialize the first element of `Nt` to be equal to `N0`, as we did above.
    You want to calculate η_t, which depends on η_{t-1}. For the first time step, there is no η_{t-1}, however. You therefore set $\eta_t = v_t$ and use that to calculate r_t for the first time step:

```
etat[1] <- vt[1]
rt[1] <- mu + etat[1]
```

Then, loop through years, each time calculating `etat[i]` and `rt[i]`:

```
n.loops <- length(years)
for (i in 2:n.loops) {
  etat[i] <- rho * etat[i-1] + sqrt(1-rho^2)* vt[i]
  rt[i] <- mu + etat[i]
}
```

The rest of the code is the same. Thus, the entire code looks like

```
years <- 0:tmax
n.loops <- length(years)
output <- rep(x = NA, times = 1000)
for (j in 1:1000) {
  Nt <- rt <- etat <- rep(x = NA, times = tmax)
  vt <- rnorm(n = length(years),  mean = mu,  sd = sigma)
  Nt[1] <- N0
  etat[1] <- vt[1]
  rt[1] <- mu + etat[1]
  # loop to calculate etat and rt
  for (i in 2:n.loops) {
    etat[i] <- rho * etat[i - 1] + sqrt(1 - rho ^ 2) * vt[i]
    rt[i] <- mu + etat[i]
  }
  # loop to run recursive equation
  for (i in 2:n.loops)  Nt[i] <- exp(rt[i-1]) * Nt[i-1]
  # evaluate model and save to object output
  minpop <- min(Nt)
  extinct <- ifelse(minpop < Nextinct, yes = 1, no = 0)
```

```
    output[j] <- extinct
}
sum(output) / 1000
```

You're all done. Just rerun your R script, and you'll get a new extinction risk estimate.

14.4.4   Propagating uncertainty

Currently, your model takes the true model parameters (μ and σ) and uses that to generate 1,000 versions of a fifty-year model run, in which the r_t are random. What if you wanted to incorporate the fact that the model parameters are not known perfectly? In chapter 13, we saw that we can also make the parameters be random draws. So, it is relatively easy to modify the spreadsheet to ensure that not only are the r_t random, but also the underlying parameters describing them.

We will need to assign a few more parameter values: one that contains the mean of μ (you might call this mu_hat) and the standard error (called "SE" and calculated as the standard deviation divided by the sample size). In both spreadsheets and R, we can generate random draws of μ by sampling a a normal distribution with a mean equal to 0.0016 and a standard deviation equal to 0.006.

We will also need to calculate a random draw of the standard deviation of r_t, based on the sample standard deviation, which we will denote $\hat{\sigma}$. The formula for this is

$$\sigma = \sqrt{\frac{\hat{\sigma}^2 c}{n-1}}$$
$$c \sim \chi^2_{df=n-1} \qquad\qquad (14.26)$$

which says that c is a random variable that follows a chi-squared distribution with degrees of freedom equal to $n - 1$, where n is the sample size (42), and $\hat{\sigma}$ is the sample standard deviation (here, equal to 0.038).

14.4.4.1   Spreadsheet guidance

We will modify the equation you generated above in section 14.4.1. First, in cell B1 (or whatever cell contains the value for μ), replace the value with a formula that will return a draw from a normal distribution:

=NORM.INV(RAND(), muhat, se)

Then generate a random draw for σ, by first getting a random draw from a chi-squared distribution:

=CHISQ.INV(RAND(), 41)

and then use that draw as the random variable c in equation 14.26. We use 41 because we had $n = 42$ data points

You can now simply rerun your macro to propagate uncertainty in the parameters and evaluate extinction risk.

14.4.4.2 R guidance

You only need to create a few new objects and add two new lines to your R code above (section 14.4.2).

First, create an object called mu.hat and assign it the value 0.0016. This is the mean of the average population growth rates. I know, it is confusing to think of a mean of the mean. Then, create an object se, which is the standard error of the mean. Finally, we need an object to hold the sample standard deviation sigma.hat: `sigma.hat <- 0.038`.

The only thing left to add is within the outer loop. Here, for each of the *j* iterations of the outer loop you will randomly draw a value of mu and `sigma` for each iteration. So, the first two lines of the outer loop would be

```
mu <- rnorm(n = 1, mean = mu.hat, sd = se)
sigma <- sqrt (sigma.hat ^2 * rschisq(n = 1, df = 41) / 41)
```

You run your code as before, and just note that you have variability entering in two different ways. First, between each iteration, the model parameters change. Second, within each iteration, the r_t change.

14.5 Numerical solutions to differential equations

For much of this book, we used discrete-time models in our examples. One reason for this is that they are much simpler to program and to run, because our spreadsheets or R scripts like having discrete-time increments. They are nice and tidy, because state variables at one time step are calculated directly from state variables at earlier time steps.

But suppose you want to use a continuous-time model? In these, time is continuous, and the state variables and the rates that depend upon them are changing constantly. But our computers like having discrete-time steps. In spreadsheets, we have discrete rows. In R, we have discrete iterations through loops.

Suppose we had the following simple model of density-independent population growth:

$$\frac{dN}{dt} = gN$$

When $g > 0$, the state variable N is constantly getting bigger and the rate of population growth is constantly growing. This particular example has an analytical solution ($N(t) = N(0)e^{gt}$). But there are many models for which we don't have this luxury. In these cases, we need to use a numerical algorithm to simulate the time dynamics of our state variable.

In all of our examples, we'll assume that $N(0)$ equals 10 and g equals 1 yr^{-1}. Note that this is a very high population growth rate, but one that is useful to illustrate the different methods. We'll compare a series of increasingly complex methods against the true value of $N(t)$.

14.5.1   The Euler method

First things first. You need to know that the Euler method doesn't work very well. So, why devote any time to it? Two reasons. First, it nicely illustrates the challenge in numerically solving differential questions and motivates the solutions that appear in better methods. Second, the method more or less underpins the other methods.

The Euler method works like this. Given any differential equation

$$\frac{dX}{dt} = f(X) \tag{14.27}$$

you approximate the dynamics by incrementally adjusting $X(t)$ over small time steps Δt. In other words, approximate the differential equation with

$$\frac{\Delta X}{\Delta t} = f(X) \tag{14.28}$$

Essentially, the Euler method says, "If the differential equation is just the limit as Δt becomes very small, why not just use very small time increments?" In other words, we'll take any differential equation and approximate it with

$$X(t + \Delta t) = X(t) + \Delta t f(X(t)) \tag{14.29}$$

for very small Δt.

How well does this perform? figure 14.2 shows results for two values of Δt: first 0.1 and then 0.05 The true value of $N(t)$ is the gray line. The lower black line is the Euler result when Δt equals 0.1, and the one slightly higher up is the Euler result when Δt equals 0.05.

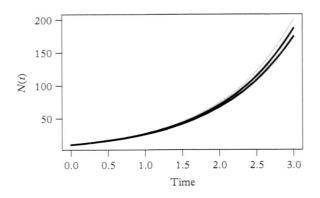

Figure 14.2 *The Euler method is not very accurate, here even with small values of Δt.*

Clearly, the Euler method doesn't work very well. In fact, to get reasonable accuracy in this example, I would have to make Δt very small, about 0.001, which means I need to do 1,000 calculations for each time step. This is pretty inefficient. And the problems get worse as time goes on. Had I run this out to twenty years, I would need even smaller time increments to be accurate.

We start with this inaccurate method because it forms the basis of all of the other methods. That is, they all try to approximate the differential equation with a discrete recursive equation, but they use different methods to predict the increment from one time step to the next.

14.5.2   The Adams-Bashford method

The Euler method is inaccurate because the value of $f(N)$, here gN, is changing rapidly within our time increments. That is, at the start of the first time interval, the population is initially increasing at a rate of g times $N = 2$ individuals yr^{-1}. By the end of the time interval, it is increasing at a faster rate. Because we use the value of $f(N)$ at the start of the time interval and apply it to the entire time interval, the Euler method was always behind the true value. The opposite would be true if $f(N)$ was slowing down over the time interval: we would consistently overestimate population size.

The Adams-Bashford method is a computationally simple method that extends the Euler method by adding one additional calculation. The Adams-Bashford method tries to figure out if $f(N)$ is accelerating or decelerating and then adjusts the predicted $N(t + \Delta t)$ accordingly.

The Adams-Bashford algorithm for any continuous model expressed as $f(X)$ is

$$X(t + \Delta t) = X(t) + \frac{\Delta t}{2}\left(3f(X(t)) - f(X(t - \Delta t))\right) \tag{14.30}$$

This might look intimidating, but let's break it down. The first part looks similar: we're simulating the differential equation in discrete-time steps defined by Δt. The second part looks quite a bit different. The term inside the bracket includes the value of the derivative calculated at that time step, minus the value of the derivative calculated in the previous time step. Essentially, it is a weighted average of the two values of $f(X)$, where the weight for the present derivative value is 3/2, and the weight for the previous derivative value is $-1/2$. This gives a better description of the average rate of change throughout the upcoming time step. Like the Euler equation, we have to multiply this average rate by Δt.

You might say, "But what do I do for the first time increment, when I don't have any value for $f(X(t - \Delta t)$?" Good question. You have to use the Euler method for the first time increment.

This is an extremely useful algorithm once you consider how you would probably set up your spreadsheet or R code. In one column, you have your values of population size at each time iteration. In the adjacent column, you have the values of $f(X)$ for each value of t. If you were using the Euler method, you would only use the $f(X)$ at the

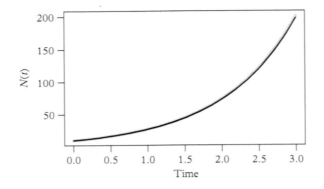

Figure 14.3 *The Adams-Bashford method improves accuracy without any additional calculations. Here the black (approximation) and gray (true) lines cannot be distinguished.*

current time step to propagate the model forward. The Adams-Bashford method also uses this calculation, plus the calculation for the previous time step, and provides a much more accurate approximation. And this comes with no extra calculations! That is, you already are calculating the function $f(X)$ at each iteration anyways, so it costs nothing to use this calculation in subsequent calculations. Thus, the Adams-Bashford method is a very efficient and simple method to implement, though not necessarily always the most accurate.

Let's see how well this performs. For our model, even a fairly large value of Δt (0.1, which is large, given the rate of change) results in very accurate predictions of population size. Compare the black line in figure 14.3 with the lower black line in figure 14.2. Both of these have the same Δt, but the Adams-Bashford method is almost 100% accurate at this large time jump. At no additional computational cost, we've nearly cracked the problem. Here we would probably want to reduce Δt a bit lower, say, to 0.05, which gets us to very nearly perfect in this example.

In summary, the Adams-Bashford method is very useful if your state variables are all changing at similar rates and do not change too fast. It is also an easy algorithm to implement on your own.

14.5.3 Runge-Kutta methods

We found that the problem with the Euler method is that it uses the value of $f(X)$ at the beginning of the time step as representative of the rate of change over the entire time step. Yet, because X is changing over the time step, the value of $f(X)$ is perhaps the least representative value of $f(X)$ that we might use for all values that X takes during that time step (the end of the time step would be equally unrepresentative).

The basic idea behind the Runge-Kutta methods is to evaluate the derivative at a number of places between t and $t + \Delta t$. As you might imagine, the midpoint method uses the halfway point between t and $t + \Delta t = t + 0.5\Delta t$. The only problem is that we don't know what value $X(t)$ will take at $t + 0.5\Delta t$. So, we use a "trial" value determined

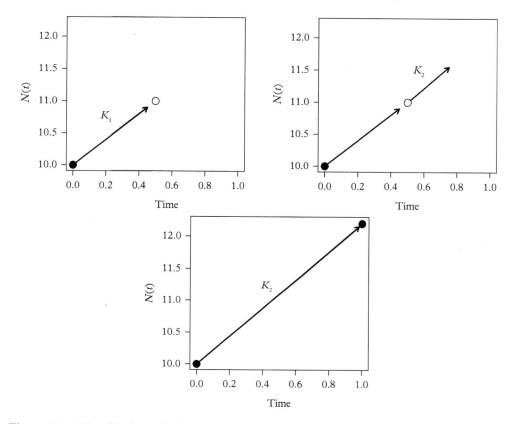

Figure 14.4 *The midpoint method uses the initial value to get an initial rate of change and then projects a trial midpoint from that (left panel). It then uses the trial midpoint to calculate the rate of change again (middle panel). It finally takes that rate of change, assumes that this is an average rate of change over the entire time period, and projects forward accordingly (right panel).*

by using Euler's method. Returning to our density-independent model, we'll consider the transition from time = 0 to time = 1 (i.e., we set Δt equal to 1).

In the simplest version, we calculate just a single trial midpoint. This trial midpoint is calculated using the Euler method to guess where $N(t)$ will be at $t = 0.5$. We'll call this first rate of change K_1, which equals $gN(0)$, or $1 \times 10 = 10$. Extrapolating this rate of change to $t = 0.5$ gives a trial midpoint of $10 + 10 \times 0.5 = 15$. The trial midpoint is illustrated as an empty circle in figure 14.4. At this trial midpoint, we calculate $f(N)$ again, this time calling the rate of change K_2. This equals $15 \times 1 = 15$. Finally, we return to the starting point $N(0)$ and project $N(1)$ as $N(0) + 15 = 25$.

This did a lot better than the Euler method, which would have predicted a population size at $N(1)$ of 20, whereas the midpoint method predicted that $N(1)$ equals 25. This compares somewhat well to the true population size, which equals 27.18. We could be even more accurate with a slightly smaller Δt (here, I used $\Delta t = 1$, which is enormous for the rate of change in this model).

Mathematically, the midpoint method is expressed with the following expressions:

$$K_1 = f(X(t))$$
$$X_{\text{trial}}(t + 0.5\Delta t) = X(t) + 0.5\Delta t K_1$$
$$K_2 = f(X_{\text{trial}})$$
$$X(t + 0.5\Delta t) = N(t) + K_2 \Delta t \qquad (14.31)$$

14.5.3.1   The fourth-order Runge-Kutta method

This method builds on the previous method but is a bit more thorough in exploring the rate of change to use over the chosen time increment. It is called "fourth order" because we will calculate four values of the rate of change for each state variable.

The fourth-order method begins the same way as the midpoint method: calculate the value of the derivative (K_1) at the initial point and derive a trial midway point. At this midpoint, calculate the value of the derivative (K_2). This is where the two methods diverge. Now you'll calculate a new estimate of the midway point by using the initial value and the derivative calculated at the midway point (i.e., K_2) (figure 14.5):

$$\text{Trial midpoint} = X(t) + 0.5\Delta t K_2 \qquad (14.32)$$

For our example, K_2 equals 10, so the trial midpoint will be $10 + 10 \times 0.5 = 15$. At this new midpoint, we will again calculate the value of the derivative, and we'll call this K_3 (here, equal to $1 \times 15 = 15$). Lastly, you will calculate a trial "endpoint" by assuming that the K_3 derivative value can be applied throughout the entire time interval:

$$\text{Trial endpoint} = X(t) = \Delta t K_3 \qquad (14.33)$$

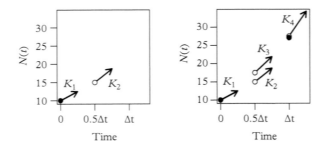

Figure 14.5 *The fourth-order model builds on the midpoint method by adding new trial points, trial endpoints, and uses a weighted average of the four rates of change that are calculated.*

Once more, we'll calculate the derivative at this trial endpoint, and we'll call this K_4. We then take a weighted average of these four estimated rates of change throughout the interval:

$$X(t + \Delta t) = X(t) + \Delta t \left(\frac{K_1}{6} + \frac{K_2}{3} + \frac{K_3}{3} + \frac{K_4}{6} \right) \qquad (14.34)$$

To summarize, K_1 is the derivative at the initial time step, K_2 is the derivative calculated at one estimate of the midpoint, K_3 is the derivative calculated at another estimate of the midpoint, and K_4 is the derivative calculated at a trial estimate of the endpoint. Thus, the terms in the brackets in equation 14.34 are a weighted average of these, where the derivatives calculated at the two midpoints are given twice as much weight as the derivatives calculated at either the initial point or the endpoint.

This algorithm is indeed very accurate, allowing you to use large time steps (Δt) and still get accurate results. Indeed, for our example, the predicted value of $N(1)$ (27.08) is very close to the true value of $N(1)$ (27.18), even though our Δt is very large.

The disadvantage is that there are several calculations to make for each time step, most of which you throw away after you've completed each step. For complex models, though, the extra calculations are more than compensated for by the ability to set bigger step sizes. There are some advanced variations on this method, though, that can make it very fast. For instance, there are algorithms that will adjust the time step width, based on how rapidly $X(t)$ is changing. If it's changing very slowly, it will use very large steps. When it's changing very rapidly, it'll compress the time steps. Most programming applications will have packages built in that will do these calculations for you automatically.

15

Sensitivity Analysis

15.1 Introduction

This chapter of Part 3 is a bit of a departure. Rather than reviewing specifics of coding, it reviews a type of model analysis—sensitivity analysis. A sensitivity analysis is an attempt to identify the parts of the model (structure, parameter values) that are most important in governing the output. It is an important part of modeling because we use it to quantify the degree of uncertainty in the model prediction, and, in many cases, it is the main goal of the model (i.e., the model was developed to identify the most important ecological processes). It also requires some careful thought about what model outputs are of interest, and what part of the model will be evaluated for sensitivity.

We've seen this in some ways already. In section 2.5, we saw that we can get different dynamic behavior by changing one parameter in the logistic population model. In section 3.4, we saw that we can identify which parameters have the biggest influence on the population growth rate in a structured population model. We saw (section 4.3) that the stability of a population model depends on the model parameters.

Here, we'll treat this type of analysis in more detail, giving some common procedures for different types of sensitivity analyses.

15.1.1 Types of sensitivity analysis

We often consider models as having two main sources of uncertainty. The first is parameter uncertainty. You likely had to specify values for parameters in your models, but some of those parameter values might not be well known. What, for example, is the per capita birth rate? You probably would like to know how the model output depends on the this value, especially if you didn't have much confidence in your estimate.

The second is structural uncertainty. Your model will have several structural assumptions as part of the model-bounding exercise (when you simplified the real-world breadth and detail). What if by doing so you missed an important structure of the real world? For example, what if a population model did not include density-dependent mortality when, in fact, this population suffers from severe population collapses whenever adult population density becomes very high? What is the sensitivity of your model to this structural assumption?

Introduction to Quantitative Ecology: Mathematical and Statistical Modelling for Beginners. Timothy E. Essington, Oxford University Press. © Timothy E. Essington 2021.
DOI: 10.1093/oso/9780192843470.003.0015

Sensitivity analysis is a term that encompasses a broad range of model analyses, all of which seek to identify important parameters that govern the model. We use sensitivity analysis to identify key uncertainties for future study, to weigh confidence in the model predictions (e.g., if the model is highly dependent on a parameter that is poorly known), and to reveal important underlying relationships. Most any model should undergo some sensitivity analysis.

15.1.2   Steps in sensitivity analysis

First you need to choose the model output that you want to evaluate. This sounds simple but is actually a bit nuanced. Most models have a large number of possible outputs. Take even a simple unstructured population model. Your output of interest might be the population size at a specific year (and even then, which year do you choose?), the change in population size from some initial state to some final state, the minimum or maximum population size, or the number of years that the population was above or below some population threshold. I'm sure you could think of some more. Your choice of model output should be guided by the question that is being asked.

Next you need to choose the kind of sensitivity analysis. Are you conducting a sensitivity analysis to explore parameter uncertainty, or are you conducting a sensitivity analysis to explore structural uncertainty? The former is much more straightforward and there are standard methods that we'll explore below. The latter, on the other hand, simply means, "Look at major structural assumptions and confirm that the result you are getting is robust to these decisions." I'll give one brief example of this later. For now, presume we are doing parameter sensitivity analysis.

Finally, one needs to use outputs of sensitivity analysis to guide exploration of the model to understand the results. Why is the model particularly sensitive to some parameters? Why is the model particularly insensitive to others? Did the sensitivity analysis reveal a problem with the model or reveal a feature of the model that was not intended? This is a key part of learning about the real world and happens by fully understanding our models (section 1.3.1)

There are two primary ways of evaluating the sensitivity of a model to its parameters. The first is called individual perturbation analysis, where you evaluate the effect of perturbing each parameter individually. The second uses Monte Carlo simulation (section 14.3) to vary all parameters simultaneously to see the distribution of model outputs and then use those to see which parameters affected that distribution the most.

15.1.3   Example: Tree snakes

We'll base our exploration of sensitivity analysis around an interesting model of tree snake control measures on the Hawaiian island of Oʻahu (Burnett et al. 2008). Brown tree snakes are an invasive species in a number of central Pacific Islands. They are perhaps most famously known for their invasion of Guam, where they now are extraordinarily abundant, and millions of dollars are spent annually to control them. At the same time, these snakes cause considerable economic and ecological damage. They predate native birds, causing the extinction of indigenous fauna. They also bite people, contributing to medical costs via emergency room visits. One of the higher economic costs derive from power outages that occur when snakes span power lines. Control measures including trapping and poisoning.

Because of the potential high costs associated with invasion, many governments have invested in snake-prevention programs. These cost quite a bit of money: annually around 2.3 million USD is spent on the Hawaiian island of Oʻahu to prevent the establishment of brown tree snakes on the island.

Choosing the right investment in different management policies might be informed by a model. Using a model, we can generate plausible biological responses of tree snakes to removal and use some empirical data relating actions to economic costs. However, any model is likely to have considerable uncertainty, so the model needs to be explored thoroughly to understand the sensitivity of the model to its assumptions.

A brief overview of the model is provided in figure 15.1, which is based on the model by Burnett et al. (2008) but modified slightly for clarity. Briefly, it is a deterministic,

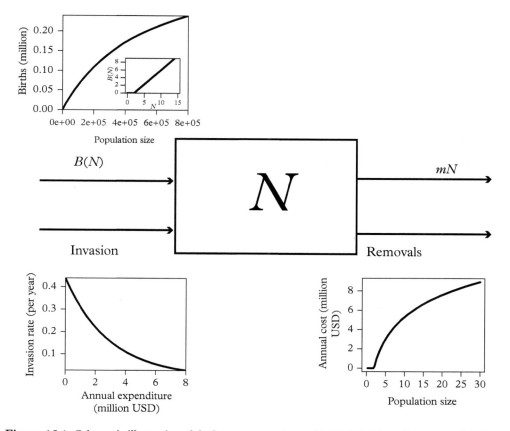

Figure 15.1 *Schematic illustration of the brown tree snake model. Modified from Burnett et al. (2008).*

unstructured, density-dependent model. There are two main policy levers: the amount to invest in inspections to prevent new invasions (call this Y_p), and the amount to invest in eradication efforts to reduce snake densities to a target snake density, here called N_{target}. Based on these choices, the dynamic model will project forward population sizes for 100 years, tallying economic and environmental costs of tree snakes, and costs of prevention and irradication.

More details on the tree snake model

The main model is the tree snake dynamic equation, where N is the number of tree snakes on Oʻahu:

$$N_{t+1} = N_t + B(N_t) + \lambda - d(N_t) - x_t \qquad (15.1)$$

where $B(N)$ is a density-dependent function depicting births as a function of population abundance; λ is the invasion rate, which depends on the expenditures to search cargo ships and other points of entry; $d(N)$ is the number of deaths, which is density independent and equal to mN_t, and x_t is the number of snakes removed each year via trapping. Per-capita birth rate equals:

$$b(N) = \begin{cases} \frac{\alpha(N - N_{min})}{1 + \beta(N - N_{min})} & \text{if } N > N_{min} \\ 0 & \text{otherwise} \end{cases} \qquad (15.2)$$

where N_{min} is the minimum population size needed for reproduction, and the parameters α and β are described in eq. (2.19).

Invasion rate declines with investment in prevention, Y_p:

$$\lambda = e^{a - bY_p} \qquad (15.3)$$

where the parameter a gives the log-invasion rate with no investment in prevention, and b is the effective reduction from each million USD investment in prevention.

I modified their model so that the cost of removing each snake depends on the level of population depletion (N_{target}) that is desired. As you try to get smaller N_{target}, the costs that are needed to reach that level increase dramatically. This is because it is relatively inexpensive to use searching and trapping to keep snake levels in the 100s, but much more money is needed to reduce snake levels to extremely low levels. The relationship between per snake removal cost y_r and target abundance is

$$y_r = -c_r \log\left(1 - \frac{(N_t - N_{target})}{N_t}\right) q^{-1}$$

$$Y_r = \max(0, N_t - N_{target}) y_r \qquad (15.4)$$

continued

where Y_r is the total cost of snake removal, c_r is the cost per unit of removal effort, and q is the fraction of snakes removed per unit of removal effort.

The third cost is the damages associated with snakes. This is presumed to be a linear function of snake density, equal to 122.31 USD annual damage per snake.

In this example, the output variable is total expenses (prevention cost, eradication cost, plus economic/environmenal damages).

15.2   Individual parameter perturbation

This is a simple method to judge the influence of one parameter on model outcome. The name comes from the following:

individual parameter: change one parameter at a time

perturbation: change the parameter value

We wish to determine how the model output changes in response to a change in the parameter value. Put another way, we want to determine how a change in some model parameter, θ_i, dictates the model result, X. Typically, we evaluate sensitivity in the neighborhood of a set of "nominal" parameter values, which are our best-guess estimates. For this reason, we'll refer to this as "local individual parameter perturbation," because we are judging the sensitivity locally, not globally.

Consider a model with four parameters (figure 15.2). The simplest measure of sensitivity is the slope of the model output with respect to each model parameter. So, if there are i parameters, you could calculate $\frac{dX}{d\theta_i}$ for each of the i parameters. For some models, you can use calculus to measure this, but, most of the time, you'll need an approximation. For example, you can change each parameter by a small amount, say, 10%, calculate the change in model output, and then approximate the derivative. Recall that, to calculate a slope, you calculate the change in X divided by the change in θ. Because this procedure reveals sensitivity near the model nominal parameter values, this provides a measure of "local" sensitivity.

Because this method is called "individual parameter perturbation," we only change one parameter at a time. So first, we ask how X responds to a small change in θ_1. We then return θ_1 to its nominal value and repeat for θ_2, θ_3, and θ_4.

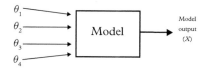

Figure 15.2 *We want to know the sensitivity of the model output (right) to the model parameters (left).*

15.2.1   Quantifying sensitivity

One problem with the derivative approximation above is that the slopes all have different units, so direct comparison is difficult. Here is an example. You wish to examine the sensitivity of population size fifteen years into the future, from a logistic model. You have two parameters: r and K; K might be numbers of animals per hectare, and r has units of y^{-1}. These are completely different units, which means the slopes are scaled completely differently and aren't directly comparable. Parameters that are normally small, like r, will, of course, have larger slopes than parameters that can be huge, like K.

One way to solve this problem is to standardize the comparison and make them unitless. We do this by measuring the proportional change in X with a given proportional change in the parameter value θ:

$$\frac{\frac{\Delta X}{X^*}}{\frac{\Delta \theta_i}{\theta_i^*}} \tag{15.5}$$

This equation looks more confusing than it is. The numerator is just the change in model output expressed as a proportion of the initial value of X (i.e., X). The denominator is the change in the model parameter expressed as a proportion of the initial value of the parameter (θ_i). Presuming that you increased each parameter by 10%, the entire denominator will simply equal 0.1, so this expression simplifies to

$$\frac{\Delta X}{0.1 X^*} \tag{15.6}$$

A value of 1 means that a change in a parameter value produces an equivalent change in model output. Typically, any values greater than 1 or less than -1 indicate *very* high sensitivity.

This language might sound familiar. That is because we have already encountered this language in section 3.4. Yes, elasticity analysis is a scaled sensitivity analysis. What we have done here is broadened the range of models where we can apply this method and created an approximation to the derivatives so that we can apply it in any model.

The advantages of this approach are that

- it is easy to implement, and
- it gives an intuitive measure of parameter sensitivity.

The disadvantages of this approach are that

- it only considers "local" sensitivity,
- sensitivity may depend on other parameter values,
- sensitivities may not be linear over larger ranges of parameter values, and

– the approach doesn't consider how well each parameter is "known" (e.g., it does not explicitly consider the uncertainty in the parameter values).

In other words, all of the advantages and disadvantages of elasticity analysis apply to local individual parameter perturbation.

15.2.2   Global sensitivity analysis

The analysis above measures "local" sensitivity, that is, how sensitive is the model to small changes in the models? Thus, the set of parameter values define the neighborhood, and the sensitivity only applies to locally near that neighborhood.

The other way to use individual perturbation analysis is to measure the model output over a wider range of parameter values. Here, there is no single estimated quantity that measures sensitivity, but instead one can visually assess how the model output changes as a parameter value changes. The typical approach is to specify reasonable upper and lower limits and then calculate the model over several possible values within these limits. By plotting the results (figure 15.3), you can identify the shape of the sensitivity (e.g., are there ranges where the sensitivity is high or low?).

For instance, the tree snake model predicts increasing predicted total cost as the parameter a increases, but the slope is steepest at small and large values of a. The increase in slope at the high end makes sense, because invasion rate is proportional to e^a, and that relationship is largely what we are seeing at the upper end of the range. When a is very small, the model is also sensitive to small changes in the parameter. This sensitivity arises because when the invasion rate very low, the population spends many years below the minimum threshold needed for reproduction to occur. Increasing the snake invasion rate in this space leads to a rapid increase in the time until reproduction occurs, which, in turn, increases total costs.

An even more complex picture is seen with the sensitivity of the model to the snake death rate m. Here there is a broad region of m above 0.14 where changing the value

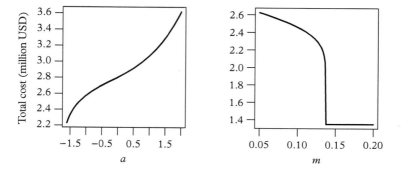

Figure 15.3 *A global sensitivity analysis looks at how the model output varies with the range of values for each parameter. The left panel shows the sensitivity to the parameter* a, *the maximum tree snake invasion rate; the right panel shows the sensitivity to the parameter* m, *the snake mortality rate.*

of *m* has no effect on the projected cost. This happens because, at the base model parameterization, *m* greater than 0.14 means that the snake population cannot sustain itself naturally. Thus, snakes that invade naturally die out, incurring very little ecological cost and no removal costs. However, there is an abrupt threshold where this shifts: when the invasion rate exceeds the rate of mortality such that the population can grow via natural reproduction and thereby increase the costs of removal.

Note that, in both of these cases, the shape of the relationship likely wouldn't have been perfectly predicted in advance. Sure, we might have had intuition that cost will go up as invasion rate goes up and that cost will go down as mortality rate goes up. But we probably would not have anticipated the nonlinear interactions among model components that lead to different sensitivities depending on how you set the initial values of these parameters.

Also, applying these analyses are really important in the describe, explain, and interpret process. By explaining the patterns of sensitivity of the model, we learn much more about the model workings, so we can explain it better and thereby give a more thoughtful interpretation about what it means in the real world.

It is possible to do this in two dimensions as well. Given two parameters *a* and *m*, you could generate a surface of model outputs and see how it depends, globally, on the combination of these two values. By combining these in multiple ways, you can identify all sorts of interactive effects of parameters. The term "interactive effects" simply means that the sensitivity of the model to one parameter depends upon the value of the other parameters. For instance, Baskett et al. (2007) examined a model relating fitness of marine organisms to larval size, and the size and spacing of marine reserve networks (figure 15.4). The sensitivity is judged by how much the surface changes as you move along one of the axis. Take the lower left panel in figure 15.4; the model is relatively insensitive to larval size when reserve width is low; the relative fitness is always low regardless of larval size. Yet, when reserve width is large, the model becomes extremely sensitive to larval size, producing large fitness at small larval sizes, intermediate fitness at intermediate sizes, and large fitness again at large larval sizes.

15.3 The Monte Carlo method

If you really wanted to identify all of the sensitivities to the model, including interactive effects of model parameters, you'd probably want to do something like the global individual parameter perturbation described above, but, instead of changing one value at a time, you'd change several parameters at a time. With a model with two or three parameters of interest, you could generate figures like those in figure 15.4. Here, the authors were interested in the fitness of a hypothetical marine fish as a function of offspring size, mortality, and the spacing and width of reserves where they would be protected from fishing. They found highly nonlinear and strong interactive effects of the model parameters on relative fitness. These effects would have been impossible to see, let alone understand, had the authors only performed individual parameter perturbation.

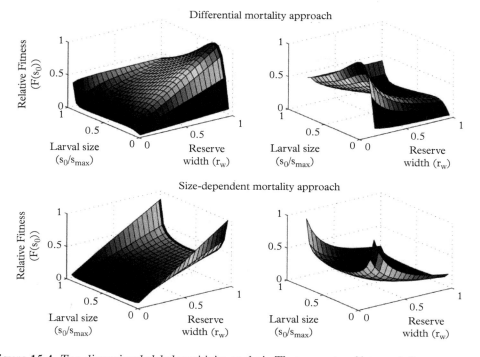

Differential mortality approach

Size-dependent mortality approach

Figure 15.4 *Two-dimensional global sensitivity analysis. The parameter of interest is fitness, measured against the size of offspring, spacing, and width of reserves, for two different assumptions about mortality in a marine fish. Figure from Baskett et al. (2007). Republished with permission of University of Chicago Press, from The American Naturalist, Marissa Baskett, Joshua Weitz and Simon Levin, Volume 170, 2007; permission conveyed through Copyright Clearance Center, Inc.*

But when you have many parameters, this approach becomes intractable, because there are too many combinations of model parameters. For example, if you had three parameters and wished to look at only twenty parameter values for each, you would have 8,000 parameter combinations to test. If you had six parameters, you would have to run 6.4 million parameter combinations to try them all. Not very efficient!

The Monte Carlo approach is an alternative way to explore parameter space to identify model sensitivities and interactions among parameters (section 14.3). The basic idea is that we specify minimum and maximum values for each parameter, randomly draw a parameter value within this range for each parameter, calculate the model output, and save the parameter combination and model output. By repeating this hundreds or thousands of times, we hope to sample the multidimensional space of plausible parameter combinations in a way that helps us identify important sensitivities and interactive effects between parameters. An interactive effect means that the sensitivity of the model to one parameter depends on the value of one or more other parameters.

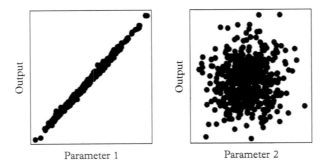

Figure 15.5 *Model depends strongly on parameter 1; parameter 2 doesn't seem to matter.*

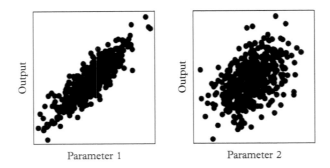

Figure 15.6 *Both parameters have an effect on the model output.*

Some characteristic outputs look like figure 15.5. We would interpret this as evidence that the model doesn't care at all about the second parameter, but the model output is hugely sensitive to the first parameter (in fact, we would be a bit worried if one parameter governed the model output so much!).

Another possibility is that multiple parameters are important, but they act in an additive fashion (figure 15.6). What we mean by additive is that parameter 1 has a certain effect on the model, and parameter 2 has a different effect on the model. The joint effect of changing both parameters is roughly the sum of the two model effects. We see that in figure 15.6. As both parameters increase, the model output increases.

In reality, these plots are often a lot more complex than those shown above. Below, I show the results of a Monte Carlo sensitivity analysis changing three variables in the brown tree snake model (the overall invasion rate, λ; the maximum rate of reproduction, α; and the mortality rate, m). The first panel in figure 15.7 looks like a scattershot of points, suggesting a lack of sensitivity of the model to invasion rate. Yet, if you look closely, you can see that there are several simulations where invasion rate is less than 0.4 and the annual cost is low and roughly equal to the prevention expenditures that the model presumes (around 1.36 million USD annually). We see a different relationship with the

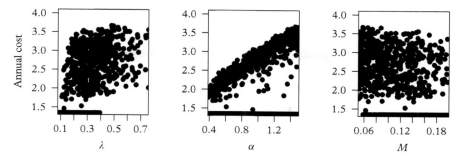

Figure 15.7 *Monte Carlo sensitivity analysis of the brown tree snake model. The resopnse variable is the total expected annual cost.*

model sensitivity to maximum reproductive rate: there is one cluster of points that seems to suggest a smooth, increasing average annual cost as reproductive rate increases. On the other hand, there are again many points positioned horizontally near cost = 1.36 million. Finally, the sensitivity to mortality is a hybrid between the other two variables: a weak but noisy declining trend in cost with increasing mortality, and again a second cluster of points positioned near 1.36 million yr_{-1}.

Findings like this are a clear indication that there are interactive effects of variables. That is, the sensitivity of the model to one variable depends on the value of one or more other variables. In this case, we likely have a three-way interaction. Some clever plotting will reveal that those points where the annual cost is near 1.36 million USD are all associated with cases where the invasion rate is insufficiently low to match the mortality rate once the snakes have invaded. In other words, there are combinations of invasion rate and mortality when the snakes will always die out on their own. Once mortality cannot keep pace with invasion rate, then you start spending a lot on removal, because snakes start reproducing naturally, and the model becomes particularly sensitive to the birth rate. We might say here that the combination of λ and mortality controls the sensitivity of the model to the birth rate parameter.

This is fairly similar to the lower left-hand panel in figure 15.4. Here reserve size was governing the sensitivity of the model to other parameters. When it was low, it negated any effect of larval size. When reserve size was large, larval size was important.

The advantages of the Monte Carlo method is that

- it directly deals with the disadvantages of the individual parameter perturbation method.
- it looks at "global" sensitivity,
- it allows you to include the uncertainty in the parameters, and
- it allows you to look at interactions between parameters.

The disadvantages of the Monte Carlo method are that

- there is no simple way to quantify and compare sensitivities, and
- it requires some programming and thoughtful analysis.

15.4   Structural uncertainty

Parameter uncertainty is reasonably straightforward (though it might involve some programming to implement). How do we address the sensitivity of our model to structural assumptions, that is, if we assume one type of relationship when in fact another exists? How can we account for that?

By far the easiest route is to transform structural uncertainty into parameter uncertainty. Take, for instance, the functional response, which is the expression that describes how the per capita rate of feeding by a predator depends on prey density, predator density, or both. The simplest, called a type 1 functional response, indicates that per capita feeding is a linear function of prey abundance. A type 2 functional response indicates says that per capita feeding is an asymptotic function of prey abundance, while a type 3 functional response indicates that per capita feeding is a sigmoidal function of prey abundance. In addition to these so-called prey-dependent functions, the function might also depend on predator abundance, to account for predator interference competition or other factors. This leads to a staggering range of possible functions you might apply (Skalski and Gilliam 2001). How might you evaluate the sensitivity of your model to the choice of function?

If you can present these alternative models as special cases when certain parameters take certain values, you can explore model sensitivity by using the usual parameter sensitivity analysis methods. For instance, I developed a model (Essington et al. 2015) of predator-prey interactions that allowed for all manner of interactions between predators and prey. The two species could be entirely independent. There could be a typical predator-prey interaction with a type 2 functional response. There could be a predator-prey interaction with a type 2 functional response with predator dependence. More interestingly, there could be complex trophic interactions whereby the prey consumes early life stages of the predator. And, depending on how the model is parameterized, it could produce an Allee effect (section 5.7.1) in the predator. All of these different model structures were represented by parameters. By setting the parameters to different values, I could toggle on and off each structure.

What if you can't do this? The standard approach is to create alternate model scenarios. That is exactly what you were seeing in figure 15.4, where Baskett et al. (2007) wanted to explore the model predictions under two very different ways of modeling mortality and how it depended on size versus life-history stage. This, of course, takes more work and analysis but is an important part of exploring model sensitivity.

16

Skills for Fitting Models to Data

This chapter will work through many of the calculations presented in Part 2 of the book. As usual, guidance is presented first for spreadsheet users and then for R users. This chapter presumes that you have already worked through the previous skills—before moving onto this section, you ought to be reasonably comfortable setting up spreadsheets and R code, and applying the modeling skills presented in chapter 14.

Attentive readers may well notice that the examples shown here are relatively simple—namely, they involve fitting parameters of probability functions, or relatively simple functional relationships. They do not yet illustrate how to combine the knowledge of dynamic modeling with that of statistical model fitting. That topic is reserved for the final chapter, chapter 18. It might be tempting to jump ahead, but please avoid that temptation. Those more advanced skills will be better learned once you master the core concepts.

16.1 Maximum likelihood estimation

In many cases (including most of the examples demonstrated here), you can estimate parameters by using mathematical solutions. For instance, you've probably estimated the mean of a normal probability distribution from data before—you simply set it equal to the sample mean. You've likely also estimated the standard deviation using sums of squares, and the sample size.

There is absolutely nothing wrong with estimating parameters that way. In fact, it's great: it is an exact solution! The problem is that, for many models, no such clever mathematical solution exists. In these cases, we ask our computers to systematically search the parameters to find those that *appear* to fit the model best.

This chapter will review two approaches. The first is to follow the guidelines given in section 8.2. The method described there is probably the best way to *learn* about maximum likelihood estimation and works fairly well for problems with small numbers of parameters. The second method, which used numerical searching algorithms, is probably the best way to *apply* maximum likelihood techniques to actual problems.

Introduction to Quantitative Ecology: Mathematical and Statistical Modelling for Beginners. Timothy E. Essington, Oxford University Press. © Timothy E. Essington 2021.
DOI: 10.1093/oso/9780192843470.003.0016

16.1.1   Maximum likelihood estimation: Direct method

In section 8.4, we saw an example where several 20 m^2 plots were surveyed, and 0, 3, 6, and 7 geoducks were seen. We want to assume a Poisson probability mass function and use likelihood to estimate the parameter r, which is the geoduck density. Here we will review the steps to estimate r and to generate a 95% confidence interval for r.

For both R and Excel, the setup will be the same. You will specify a range of different values of r, you will specify the outcomes, and then you will calculate the negative log-likelihood for each value of r. The value of r that has the smallest negative log-likelihood is the maximum likelihood estimate. The smallest and largest values of r that have a negative log-likelihood less than or equal to the smallest negative log-likelihood plus 1.92 are your lower and upper bounds, respectively, of the confidence interval.

16.1.1.1   Excel guidance

1. Name the cell that contains the value for the parameter r (call it "r_poisson" [this is because spreadsheets will not allow you to use the name "r"]). Assign a starting value.

2. On your spreadsheet, use the POISSON.DIST() function to get the likelihood for each observation, based on whatever value is listed in your named cell r_pois. The syntax would be =POISSON.DIST(**cell reference, r_poisson * area searched, FALSE**). The last statement, "FALSE," means that you want to return the probability mass not the cumulative probability. The cell reference will be where one of the observations are listed. It may look like table 16.1. You then copy that formula down the column so that you calculate the likelihood for each observation.

3. Calculate the negative-log of each likelihood using **= - LN(cell reference)**, where cell reference contains a likelihood value (e.g., cell D4 in table 16.1).

4. In some empty cell, say cell D2, write a formula that adds together all of the negative log-likelihoods (e.g., **=SUM(D4:D7)**) and name the cell, if you wish.

Table 16.1 *Setting up your worksheet for maximum likelihood estimation*

	A	B	C	D
1	r_poisson	0.05		
2				
3	Area	Number Seen	Likelihood	Negative Log-Likelihood
4	20	0	=POISSON.DIST(B4, r_poisson * A4, FALSE)	= - LN(C4)
5	20	3	...	...
6	20	6	...	...
7	20	7	...	...

Now you have a spreadsheet that takes a single value of *r* and returns the negative log-likelihood associated with that value of *r*. The next step is to use the data table (section 12.4) to make Excel loop through different candidate values of *r* and record the negative log-likelihood for each.

You might choose to put the data table over in columns F and G. In column F, starting in row 4, list values of *r* starting at 0.02, in increments of 0.01, up to *r* = 0.40. Then, in cell G3, write a cell reference that will return the negative log-likelihood (if you used the setup described above, this would be **=D2**). Select cells from F3 to G42, open the data table window, and enter as the column input cell r_poisson (presuming you named the cell B1 as described above). Excel will now automatically return the negative log-likelihood for each value of *r*.

Your final steps are relatively simple. First, find the value of *r* that has the smallest negative log-likelihood. You should find that the negative log-likelihood is minimized (at 10.715) when *r* = 0.2. This is your maximum likelihood estimate. We add 1.92 to this value to find the negative log-likelihood threshold that defines the lower and upper bound of our confidence intervals; 10.715 plus 1.92 gives 12.63. Starting at the maximum likelihood value of *r*, begin scanning upwards on your spreadsheet to find the smallest value of *r* that has a negative log-likelihood that is 12.63 or smaller. That is your lower bound. Do the same to find the upper bound. I get 0.12 as my lower bound, and 0.31 as my upper bound.

16.1.1.2 R guidance

First, create the data:

```
thedata <- c(0,3,6,7)
t <- 20
```

Then, create a function that will calculate the negative log-likelihood. We could do this by looping through observations:

```
nll.fun <- function(r, thedata, t) {
  n.data <- length(thedata)
  nll <- rep(x = NA, times = n.data)
  for (i in 1:n.data) {
    nll[i] <- - log(dpois(x = thedata[i], lambda = r * t))
  }
  return(sum(nll))
}
```

This loops through the observations, calculates the likelihood, takes the log, and then multiplies by −1. It then adds them together and returns the sum.

Normally, I avoid showing "look how tidy this code can be" because, at this point, your focus should be on code that both works and is easily understandable. But, in this case, I think it is important to see a more condensed version, because you're likely to encounter it in other sources. Specifically, you can take advantage of the fact that the R function dpois, the function that returns the probability for a given observation, can handle vectors of data and return the probability of all of them. So, our function above can be written more compactly as

```
nll.fun <- function(r, thedata, t)
  -sum(dpois(x = thedata, lambda = r*t, log = TRUE))
```

Note a few things I did here that seem a bit unusual. First, I used the option log = TRUE to have dpois return the log-likelihood. Second, rather than looping, I sent the entire vector of thedata to the function call. This returns a vector of the log-likelihood of each observation. I then put sum() around the dpois call, which means R will sum all of the log-likelihoods. Putting a minus sign in front makes this the negative log-likelihood.

Now that I have a function that can take the parameter value, thedata, and the area searched, I can call that function for many different values of *r*, and I can calculate the negative log-likelihood for each:

```
r.list <- seq(from = 0.02, to = 0.4, by = 0.01)
nll.list <- rep(x = NA, times = length(r.list))
for (i in 1:length(r.list)){
  nll.list[i] <- nll.fun(r.list[i], thedata, t)
}
```

This code creates a list of values of *r* from 0.02 to 0.4 in increments of 0.01, and, for each, calculates the negative log-likelihood and places it in an array called nll.list. This is also a good time to show one of the apply functions, as the entire loop can be replaced by

```
nll.list <- sapply(X = r.list, FUN = nll.fun,thedata, t)
```

Now we simply need to find the value of *r* that has the lowest negative log-likelihood. An easy way to do this is visually:

```
View(cbind(r.list, nll.list))
```

Or if you want to use code, you can use

```
min.nll <- min(nll.list)
r.mle <- r.list[nll.list == min.nll]
```

where the first line is finding the smallest value of the negative log-likelihood, and the second line is finding the value of the array r.list that has a negative log-likelihood equal to this value.

You can similarly find the lower and upper bounds like this:

```
target.nll <- min.nll + 1.92
ci.index <- which(nll.list <=target.nll) #which are below
  threshold
lb <- r.list[min(ci.index)] #smallest value below threshold
ub <- r.list[max(ci.index)] #largest value below threshold
```

where `lb` and `ub` are the lower and upper bounds, equal to 0.12 and 0.31, respectively.

16.1.1.3 *More precise confidence interval bounds*

Here you might notice that the selection of upper and lower bounds is a bit unsatisfying because, for instance, $r = 0.33$ was just barely above the negative log-likelihood threshold. You're right—this is a very rough way to approximate the confidence intervals. Later, in section 16.3, I will show you how to linearly interpolate to estimate the value of r that has a negative log-likelihood exactly equal to the threshold value.

16.1.2   Maximum likelihood estimation: Numerical methods

This section illustrates the application of numerical methods to yield more precise (usually) estimates of maximum likelihood parameter estimates.

No matter what platform you use, parameter estimation using numerical methods requires the same basic steps:

1. You must assign initial values for the parameters that you are interested in. These are rough guesses that you use to initialize the searching process. If you have no idea what might be a good guess, then put in values that are plausible.
2. You need to either use your spreadsheet or write a R function to take a single set of parameter values and return the negative log-likelihood. Fortunately, that is what we did above.
3. You must apply a numerical method to adjust the model parameters to make the negative log-likelihood as small as possible.

It is also worth knowing how the computer algorithms work. Essentially, they search through model parameter values, trying to find those that optimize some function (i.e., they minimize the negative log-likelihood or maximize the log-likelihood). They often track "gradients," meaning they detect things like "if I move the parameter this way, the likelihood improves" and then try to follow those gradients to hone in on the best model parameter values. There are two main ways that it does this. One is by calculating the derivative of the likelihood function with respect to the model parameters. Not surprisingly, these are called "derivative-based methods." The other is by trial and error and then expanding and contracting the range of parameter values that are being attempted. These are called "Simplex" methods.

The other thing to know is that these methods are fallible. Just because your computer landed on a value, you should not take it for granted that the value is valid; you should always investigate to confirm that it is a reasonable answer. And you shouldn't be too surprised if the model estimates are occasionally nonsense. Sometimes these gradients can lead the algorithms astray, and the ability to find the true optimal solution depends on the shape of the function that is being optimized. In general, optimization routines perform far better if they work on log-likelihoods rather than raw likelihoods.

Getting around parameter bounds through link functions

You will often want to estimate parameters that you already know must lie in some range. For instance, the density parameter in a Poisson has to be greater than or equal to 0 (presuming that the data contained any observations). Some of the routines allow you to set these limits, called "bounds," but I always seem to have more trouble with bounded optimization than with unbounded optimization. For that reason, I try to set up my optimization problem so that I don't have to bound my models. Here are two common so-called link functions that you can use to link the parameter that you are estimating to another parameter that your spreadsheet or R adjusts:

– log link: You define a parameter, say, log_r, to be the logarithm of the Poisson density. The spreadsheet or R function will then take that parameter value to calculate r as =exp(log_r). You can then solve by changing log_r. In this way, r will never become negative.
– logit link: In the binomial probability density function, the parameter p has to be between 0 and 1. You can define a parameter for estimation as p_logit. Then, the spreadsheet or R function will take that parameter, and calculate p as $= 1/(1 + e^{-p\_logit})$.

16.1.2.1   Spreadsheet guidance

Most installations of Microsoft Excel include Solver, a built-in set of optimization routines. All of your modeling work will be done in Excel, and you'll call Solver to iteratively change parameter values to solve your problem. The base installation of Excel often won't immediately enable Solver, so you'll need to activate it yourself. The steps change all of the time as Excel evolves, so you may need to do an internet search. Here are some quick instructions that work at the time of writing.

PC: Go to the Excel options window, and select Add-ins on the list on the left hand side. Over on the corresponding frame on the right, look for the choice on the bottom of the frame called Manage Add-Ins, and click "Go." A dialogue box should pop up that includes Solver as an option. If so, just click the box next to the word "solver" and then click OK and you're done. If Solver is not listed, click on the "select" button and search for "solver.xlam" to proceed.

Mac: Go the \Tools main menu and then select Add-Ins. That will pull up the same dialogue box as is described for the PC version.

Assume we are working on the example posed in section 8.4, where 0, 3, 6 and 7 geoducks were observed in four separate 20 m² plots. You wish to use solver to find the maximum likelihood estimate of the Poisson parameter r.

First, set up your spreadsheet as described in section 16.1.1.1. Then, open up the Solver dialogue box. Solver usually appears on the Data tab within the ribbon. On a Mac, it also appears under the main menu \Tools. Once you find it, select Solver. Once there, a dialogue box will appear. Here is a brief review of what it is asking:

- The prompt "Set Objective" is asking, "What cell contains the value that you want me to optimize?" Here it is the cell containing the negative log-likelihood (D2, in our example above). If you named that cell, you could type in the name here, or you can enter the cell address.

- The prompt "To" is asking, "Do you want Solver to make the objective as big as possible, as small as possible, or equal to some specified value?" We want to minimize, so select Min.

- "By changing variable cells" is where you tell Solver what values to adjust to optimize your objective cell. Here, enter the cell that contains the value for r_poisson (or if it is named, you can type in the name).

- "Subject to constraints" is a place where you can put bounds on the parameters. I try to avoid these when possible, but sometimes they are useful. Skip this for now.

- "Make unconstrained Variables non-negative" is very useful. If you haven't put a constraint on a parameter, you can tell Solver that parameters should not be less than 0. Usually, we like to keep that box checked when we have parameters that should always be 0 or greater.

- Finally, we need to choose an optimization method. The default is a derivative-based method which works well most of the time. You can improve by selecting the "Options" button. On the tab for "All Methods," make sure the box that says "Use Automatic Scaling" is turned on. This really helps the routine run better.

You should be all set. Click solve and hold your breath!

Solver will churn for a second or so and then show a dialogue box that asks whether you want to accept Solver's answer. If it looks as though it worked (i.e., there are no funny "#NUM!" errors on your page), then accept. If it didn't work, you will have to play around with the starting values and try again. Or try a different Solver algorithm (e.g., try the Simplex method). If it still isn't working, you might need to double-check a few things. First, are you accidentally *maximizing* the negative log-likelihood? If so, make sure that "minimize" is selected in the Solver dialogue box. Second, is Solver trying parameter values that don't make sense? For instance, is it trying to make r equal to 0? If so, you'll have to put bounds on the parameter or use link functions so that r will always exceed 0.

This was a simple example with a single parameter, but the same steps will work for any number of parameters. The only difference is that when you select "By changing variable cells," you'll select multiple cells, each one containing a parameter value that you want Solver to optimize.

Suppose we wanted to fit a negative binomial probability function to these data, using Solver. In Excel,we have one problem: the built-in negative binomial function is based on a completely different parameterization than the one we like to use. This is not a problem, because we can create our own function (section 12.5). Below is a Visual Basic script that you can use by copying it into a blank module. It creates a function `lognegbinom`, which returns the log-likehood given an observation, an expected value ($r \times t$ = lambda), and k:

```
Function lognegbinom(x, lambda, k)
Application.Volatile
logfactx = Log(Application.WorksheetFunction.Fact(x))
loggammaxplusk = Application.WorksheetFunction.GammaLn(x + k)
loggammak = Application.WorksheetFunction.GammaLn(k)
lognegbinom = loggammaxplusk - loggammak - logfactx +
  x * Log(lambda / (k + lambda)) + k * Log(k / (k + lambda))
End Function
```

Alternatively, you can write fairly complex cell formulas to do the same calculation.

Once you do that, you'll then create a named cell r_nb (if you haven't already) in cell B1, which contains the parameter r of the negative binomial, and another named cell in cell B2, which contains the parameter k, and calculate the likelihood of each observation, recognizing that the expected value is $r \times t$ (area searched), just as you did above. To calculate the negative log-likelihood in cell D4, you type in

= - lognegbinom(B4, r_nb * A4, k)

which you then copy down into cells D5 through D7. You calculate the sum of the negative log-likelihoods as you did above.

When selecting Solver, you follow the same steps. The only distinction now is that you will ask Solver to change two parameters: *r_nb* and *k*. In the Solver dialogue box, you can type in the cell names separated by a comma or select the range of cells that contains the parameter values (here, B1:B2). Remember, starting values really matter, so give it a decent guess: we already know the density should be around 0.2, so use that as your starting value. Type in something between 5 and 20 as a starting value of *k*. When I used these, Solver quickly converged to r_nb =0.2 and $k = 2$.

Regardless of how complex your model is, you'll use the same basic steps as described here: define parameters, calculate likelihoods, and then use Solver to minimize the negative log-likelihood. You'll be amazed at how well Solver can find solutions. Still, I should be honest: I rarely use Solver when I'm fitting complex models to data. Once the models become more complex, you need to hand-hold Solver more to ensure that it can fit. Plus, it is far easier to create an easily reproduced statistical analysis in a programming language than in a spreadsheet. That said, I find that learning about model fitting is far easier in spreadsheets, because you can see all of the intermediate steps that are harder to see in a programming language.

16.1.2.2 R guidance

There are many different packages and function calls available to you: `mle`, `mle2`, `bbmle`, ..., and probably more. Here I'll show how to optimize using R's base function `optim`. The basic steps in optimization in R are the same as in any platform:

1. Initialize starting values.
2. Have a function that will take as input the parameters that are to be optimized and the data and then return the negative log-likelihood.
3. Call the optimization function, identifying the starting values, the data, and the negative log-likelihood function.

If you were estimating the Poisson parameter r, and the data were the same as in section 8.4, where 0, 3, 6, and 7 geoducks were observed in four separate 20 m^2 plots, you would first create the data objects and the negative log-likelihood function as described in section 16.1.1.

Second, you create an initial guess, or starting values, for the parameters that are being estimated.

```
start.pars <- 0.2
```

Third, you run `optim` and assign the solution to some variable, here called `poisson.soln`:

```
poisson.soln <- optim(par = start.pars,
                      fn = nll.fun,
                      thedata = thedata,
                      t = t,
                      method = "Brent",
                      lower = 0,
                      upper = 10)
```

The first input into the `optim` function is `par =`, which is where you place your vector (or, in this case, a single value) of starting parameter values, and `fn` is asking for the function that will calculate the negative log-likelihood. After `fn = ...`, you can then send any other arguments that your function needs. Our function needs two things to be sent to it: `thedata` and `t`. The optimization method is "Brent," which is the only option when the model contains only a single parameter. This method requires that you specify lower and upper bounds. Here I chose 0, for obvious reasons, and then arbitrarily selected 10 because I knew there was no way the value could be bigger than this. The solution looks like this:

```
> print(poisson.soln)
$par
[1] 0.2

$value
[1] 10.71546

$counts
function gradient
      NA        NA

$convergence
[1] 0

$message
NULL
```

The variable "par" is the maximum likelihood estimate of r, and the variable "value" is the negative log-likelihood. The other outputs don't apply to the Brent function, so we can ignore them.

You can apply the same basic steps to fit a more complex model. Here, we will use the same data and fit a negative binomial function to them. You will have to fit both r and k, so we need to make a slight modification to the first input argument in our function. Specifically, your nll.function might look like this:

```
nll.fun <- function(pars, thedata, t) {
  r <- pars[1]
  k <- pars[2]
  nll <- -sum(dnbinom(thedata, size = k, mu = r*t, log = T))
  return(nll)
}
```

Note here that the first argument is a vector that contains r in the first element, and k in the second. In general, this is a useful way to structure all problems where there is more than one parameter being estimated.

Now set the starting values. A reasonable thing might be

```
start.pars <- c(0.1, 10)
```

which is starting at $r = 0.1$ and $k = 10$.

Finally, we must choose a optimization method. If we want a derivative based method, we can choose "BFGS." If we want the Simplex method, we can choose "Nelder-Mead." Here is an example using Simplex:

```
nb.soln <- optim(par = start.pars,
                 fn = nll.fun,
                 thedata = thedata,
                 t = t,
                 method = "Nelder-Mead")
```

and here is the fit:

```
> print((nb.soln$par))
[1]  0.2000024 1.9963713
```

which is nearly identical to the fit with the derivative-based BFGS method:

```
> print(nb.soln$par))
[1]  0.1999956 1.9968721
```

One last note here. The above nll.fun accepts a vector of *r* and *k*, and then optim will adjust those to minimize the negative log-likelihood. However, optim doesn't know that *r* and *k* have to be greater than 0. So you have two choices. One, you can apply "bounded optimization," where you supply upper and lower bounds on the parameters. Two, you can use a link function as described in section 16.1. For instance, the vector pars could be the log of *r* and *k*. In that case, your code might be:

```
nll.fun <- function(pars, thedata, t) {
  r <- exp(pars[1])
  k <- exp(pars[2])
  nll <- -sum(dnbinom(thedata, size = k, mu = r*t, log = T))
  return(nll)
}
```

This way, no matter what value "pars" takes, exp(pars) will always be greater than zero.

16.2   Estimating parameters that do not appear in probability functions

The above examples involved fitting parameters that directly appeared in the probability mass functions of the Poisson or the negative binomial. Earlier, we learned that we can estimate any model parameter by relating it to a parameter that does appear in the probability function. For instance, in section 8.6.2, we reviewed a model where the survivorship rate of gobies was density dependent. As a reminder, the model was

$$\mathbb{E}[N_{\text{survive}}] = N_0 p(N_0) \tag{16.1}$$
$$p(N_0) = p_{\max} e^{-\alpha N_0}$$

We wish to estimate α and $p_{\max}$.

In Excel, you'd set up the worksheet as shown in table 16.2.

Here, for each experiment, you are calculating the value of the binomial parameter p based on the initial number of gobies and the model parameters. In the adjacent cell, you are calculating the negative log-likelihood of the observation, given the number of initial gobies present and the estimated p.

As you did before, create a cell that sums the negative log-likelihood of all observations, say, in cell D2:

= SUM(D4:D18)

Table 16.2 *Setting up your worksheet to estimate goby survivorship parameters*

	A	B	C	D
1	p_max	0.2		
2	alpha	0.2		
3	Initial Number	Number Survived	p	Negative Log-Likelihood
4	6	4	=p_max * exp (- A4 * alpha)	=-LN(BINOM.DIST (B6,A6,C6,FALSE))
5	10	4	...	...
6	12	4	...	...
7	20	7	...	...
8	...	...	...	...

And, finally, you use Solver as you did before, minimizing the value in cell D2 by adjusting the cells that contain your parameter values.

In R, you'd do something like this:

```
n.init <- c(6, 10, 12, 20, 29, 43, 59, 91, 22, 25,
     8, 12, 15, 42, 68)

n.final <- c(4, 4, 4, 7,  9, 8,  18, 11, 9, 8,
   3, 7, 5, 17, 16)

nll.fun <- function(par, n.init, n.final) {
```

```
  p.max <- par[1]
  alpha <- par[2]
  ps <- p.max * exp(-alpha * n.init)
  return(-sum(dbinom(x = n.final, size = n.init, prob = ps,
    log = T)))
}
start.par <- c(0.7, 0.01)
fit <- optim(par = start.par,
             fn = nll.fun,
             n.init = n.init,
             n.final = n.final,
             method = "Nelder-Mead")
```

As before, the nll.fun is accepting the parameter array called par, extracting out the parameters from the array, calculating an array of binomial probabilities (ps) by applying the model equation, and then calculating the sum of the negative log-likelihoods, given the observations and the array of ps. Finally, use optim with some starting values of the parameters and pass along the data to the function.

16.3   Likelihood profiles

In section 8.3, we learned that we can calculate likelihoods (or, equivalently, negative log-likelihoods) for a range of alternative parameter values and then use the ratio of likelihoods (or the difference in negative log-likelihoods) to calculate confidence intervals. When there is only one parameter of interest, this is fairly intuitive to do. In a spreadsheet, you would just use the Data Table feature to repeat calculations of negative log-likelihood for a range of specified parameter values. In R, you would loop through alternative values of the parameter and, for each, save the negative log-likelihood.

When there are two or more parameters, one needs to calculate likelihood profiles (section 8.5). Recall that a likelihood profile for a parameter, ψ, involves setting the parameter ψ to specified values and, for each, adjusting the remaining model parameters to maximize the likelihood.

For models with only two parameters, we can use an approximation method that assumes that the nuisance parameter, say, k of a negative binomial distribution can only take one of several alternative discrete values. In that way, we can then calculate the negative log-likelihood for all combinations of the parameter of interest (e.g., r of the negative binomial) and the alternative values of the nuisance parameter and then identify the smallest negative log-likelihood over all of the alternative nuisance parameter values.

This method is approximate in two ways. First, the nuisance parameter likely can take any value, not discrete alternatives. So, there might be better values of the alternative parameter value that was not part of our set, and that might affect our negative log-likelihood outcomes. Second, to identify lower and upper bounds of the confidence interval, we need to identify the value of the parameter of interest whose profile negative log-likelihood value is exactly 1.92 units greater than the smallest possible negative log-likelihood. It is exceedingly unlikely that we will, by chance, have included the unique parameter value combinations that produce exactly this outcome. Instead, we find the parameter values that get us reasonably close to this cutoff.

So, why use the approximation method? Frankly, it is the best way to learn and understand what likelihood profiles are all about. Once you've mastered this, you can explore approaches that relax these approximations. This chapter will illustrate one of these—numerically finding the smallest possible negative log-likelihood for each value of r. In section 16.3.2, I illustrate how to find the parameter value whose profile negative log-likelihood is exactly 1.92 units above the minimum.

16.3.1   Profiles in spreadsheets

Start by setting up your spreadsheet to define parameters of the negative binomial distribution, r and k, use the data described in section 8.5, and calculate the total negative log-likelihood of the data. Your spreadsheet should look like the one you created in section 16.1.2. Recall you will have to add the Visual Basic function (or create a fairly complex cell formula) to calculate the negative binomial likelihood (section 16.1).

You want to calculate the negative log-likelihood for many combinations of r and k. That means you must define the parameter values that you'll consider for the parameter of interest, r, and the nuisance parameter, k. Keep in mind that you face a trade-off here. The more closely spaced you make the alternative parameter values, the more accurate your confidence interval estimate will be. But, to make parameter values more closely spaced, you need a lot more alternative parameter values, which makes the size of the problem grow.

Here suppose you are interested in values of r ranging from 0.03 to 0.30, in increments of 0.0025. Place those in a single column, starting in cell G6. We want our spreadsheet to calculate the profile negative log-likelihood for each of these.

Now we need to specify the alternative values of the nuisance parameter. I happen to know (section 8.5) that the maximum likelihood estimate of k is relatively small (less than 10), so it is reasonable to assign values of k from 0.5 to 20 in increments of 0.5. In real-life applications, it is not uncommon to start with a guess and then adjust if you have missed the important values of k. Starting in cell H5, list the values 0.5, 1.0, ..., 20 in a single row.

In cell G5 (above the list of r and to the left of the list of k), insert a formula that will return the negative log-likelihood. For instance, if your cell D2 contains the calculated negative log-likelihood, in cell G5 you would use **=D2**.

Now all you need to do is use the Data Table (section 12.4) to make your spreadsheet calculate the negative log-likelihood for each combination of r and k that appears in your table. This may take your computer some time to finish: the custom negative binomial function is rather slow, and your computer has a lot of parameter combinations to try.

Once that is done, you need to calculate the profile negative log-likelihood for *each* value of r. This, by definition, is the smallest negative log-likelihood calculated over all of the different values of k. In cell F6, you might therefore use a formula =MIN(H6:AU6). Copy this formula down the column to cell F114 and you've done it—you've calculated a likelihood profile for r.

Finding the lower and upper bounds of the confidence interval is relatively straightforward at this point. In section 16.1, you found the smallest negative log-likelihood that can be found for all r and n. Add 1.92 to that to get the threshold value that defines which values of r are within the confidence bounds. You simply need to read through the profile negative log-likelihoods, identify the smallest value of r whose profile negative log-likelihood is less than or equal to the threshold, and then do the same to find the largest value of r. I get (0.0525, 0.2225) as my confidence interval.

How could you make this less approximate and more precise? For one, you might try to linearly interpolate to estimate the value of r that would produce a profile negative log-likelihood that is exactly 1.92 units greater than the smallest negative log-likelihood.

First, create a cell that contains the threshold value of negative log-likelihood that you calculated above, and name the cell nll_target. Look through the list of profile negative log-likelihoods and find the first case where the value is less than the threshold. On my spreadsheet, it is in row 15. We want to interpolate in between this value of r and the next smallest value of r. To do this, in column E, on the row you noted above, type in the formula

=FORECAST.LINEAR(nll_target, F14:F15,E14:E15)

This tries to find the value of r, listed in column E, that will make the profile negative log-likelihood, listed in column F, exactly equal to nll_target.

You repeat this to find the upper bound. That is, find the largest value of r whose profile negative log-likelihood is less than the target value, and interpolate between that value in the next highest value of r. When I do this, I get a confidence interval of 0.0522, 0.225 (rounded to the nearest 0.001).

Second, you might try to see if there are better values of k to use than those you listed, by using numerical optimization methods. Here, you would hold r steady at one of the candidate values and then use Solver to find the value of k that minimizes the negative log-likelihood when r is held constant. You would then repeat that for all of the listed values of r. While you might be willing to do that for a handful of different values of r near the boundary, you certainly would not want to try that for all 109 different values of r on your spreadsheet. And it is not particularly easy to automate this process in spreadsheets. This is one of the tasks that is better handled in a programming language!

16.3.2　Profiles in R

Start by setting up your R script in the same way that you did in section 16.1.2 for the negative binomial fit, but use the example data in section 8.5. It should create an object to hold the observations (call these x), and the area searched (call these t). Create a function that takes as inputs a two-dimensional vector called pars (whose first element is r and the second element is k), the vector of observations, and the vector of area searched and then returns the negative log-likelihood from a negative binomial likelihood.

Use optim to find the maximum likelihood parameter estimates, and also save the accompanying negative log-likelihood. Note that you may get warning messages saying "NaNs produced"; this likely means that optim tried to use values for r or k that were less than or equal to 0. You can avoid that by using the link-function method described in section 16.1.

Now you will create another function that looks much like the one above but instead takes four inputs: the parameter *k*, the observations, the sample areas, and the value for *r*. It might look like:

```
r.nb.profile <- function (k, x, t, r) {
  nll <- rep(x = NA, times = length(k))
  for (i in 1:length(k))
    nll[i] <- - dnbinom(x = x[i], size = k, mu = t[i] * r,
      log = TRUE)
  return(sum(nll))
}
```

or, more compactly, as

```
r.nb.profile <- function (k, x, t, r)
-sum(dnbinom(x = x, size = k, mu = t*r, log = TRUE))
```

You want to calculate the negative log-likelihood for many combinations of *r* and *k*. That means you must define the parameter values that you'll consider for the parameter of interest, *r*, and the nuisance parameter *k*. Keep in mind that you face a trade-off here. The more closely spaced you make the alternative parameter values, the more accurate your confidence interval estimate will be. But, to make parameter values more closely spaced, you need a lot more alternative parameter values, which makes the size of the problem grow.

Here, suppose you are interested in values of *r* ranging from 0.025 to 0.30, in increments of 0.0025. Create a vector called rlist that contains these.

Now we need to specify the alternative values of the nuisance parameter. I happen to know (section 8.5) that the maximum likelihood estimate of *k* is relatively small (less than 10), so it is reasonable to assign values of *k* from 0.5 to 20 in increments of 0.5. In real-life applications, it is not uncommon to start with a guess and then adjust if you have missed the important values of *k*. Create a vector called klist that contains these alternative values of *k*.

Create a blank output matrix to hold the output; it should have as many rows as there are elements in rlist, and as many columns as elements in klist.

Loop through all values of rlist and klist, and, for each, call the function r.nb.profile and save the resulting negative log-likelihood:

```
nll <- matrix(data = NA, nrow = length(rlist), ncol =
  length(klist))
for (i in 1:length(rlist)){
  r<- rlist[i]
  for (j in 1:length(klist)){
    k<-klist[j]
    nll[i,j]<- r.nb.profile(k = k, x = x, t = t, r = r)
  }
}
```

To get the profile negative log-likelihood for each value of r, we just look across the columns of the matrix `nll` and identify and save the smallest negative log-likelihood. In R, you might do this by using the `apply` function:

```
r.profile <- apply(X =nll, MARGIN = 1, FUN = min)
```

This says to use the matrix called `nll` and apply the function `min` over rows (MARGIN = 1).

Finding the lower and upper bounds of the confidence interval is relatively straightforward at this point. Above, you found the smallest negative log-likelihood that can be found for all r and k. Add 1.92 to that, which gives the threshold value for defining which values of r are within the confidence bounds. You simply need to read through the profile negative log-likelihoods and identify the smallest value of r whose profile negative log-likelihood is less than or equal to the threshold, and then do the same to find the largest value of r. I get $(0.0525, 0.2225)$ as my confidence interval.

How could you make this less approximate and more precise? For one, you might try to linearly interpolate to estimate the value of r that would produce a profile negative log-likelihood that is exactly 1.92 units greater than the smallest negative log-likelihood. I find the lower bound by first noting where in the vector `r.profile` the value first switches from being greater than the threshold to being equal to the threshold. I find it happens between the eleventh and the twelfth element of the vector. I then apply the following code to interpolate:

```
target.nll <- min.nll + 1.92
nll.test <- r.profile[11:12]
r.test <- rlist[11:12]
lower <- approx(x = nll.test, y = r.test, xout = target.nll)$y
```

The object `lower` contains the estimated lower bound, estimated using the `approx` function. I do the same for the upper bound after noting that the switch happens between the eightieth and the eighty-first element:

```
nll.test <- r.profile[80:81]
r.test <- rlist[80:81]
upper <- approx(x = nll.test, y = r.test, xout = target.nll)$y
```

This gives me 0.052 to 0.225 (rounding to the nearest 0.001).

Second, you might try to see if there are better values of k to use than those you listed, by using numerical optimization methods. Here you would hold r steady at one of the candidate values and then use optim to find the value of k that minimizes the negative log-likelihood when r is held constant. This is relatively easy to do by changing your code above to be

```
start.pars <- 5
for (i in 1:length(rlist)) {
  r.2.use <- rlist[i]
  soln.profile <- optim(par = start.pars,
                        fn = r.nb.profile,
                        method = "Brent",
                        lower = 0.05,
                        upper = 20,
                        x = x,
                        t = t,
                        r = r.2.use)
  r.profile[i] <- soln.profile$value

}
```

If I interpolate to get the upper and lower bounds, I get nearly identical results as before: 0.052 and 0.226. The only difference between this and the approximation above is a slight increase of 0.001 in the upper bound. The relatively good performance of the approximation (when we treated k as discrete alternatives) is likely due to the fact that I had correctly identified the range of k to use.

Part IV

Putting It All Together and Next Steps

17

Putting It Together: Fitting a Dynamic Model

So far, you have learned about different types of dynamic models and that population models can be density independent or density dependent (section 2.1). You have learned that density-dependent models can display complex behavior such as dampened oscillations, limit cycles, and deterministic chaos (section 2.5). Furthermore, you learned that one can predict whether a model exhibits smooth growth to carrying capacity or one of the more complex behaviors, based on the slope of the recursive equation at carrying capacity (section 2.6.2). You learned that models can be either deterministic or stochastic (figure 5.1).

You also learned that likelihood is a way to estimate parameters (chapter 8). You learned that one can fit dynamic population models in one of two ways: assume all randomness is in the observation process, or assume that all randomness is in the population dynamic process (section 8.7). You learned that one can generate a confidence interval for parameters using likelihood profiles (section 8.5) and, finally, that you can use AIC to judge the weight of support for alternative hypotheses posed as mathematical models (chapter 9).

This chapter will integrate all of these topics into one exercise. Here is the context. Gray wolves (*Canis lupus*) in Washington State have recently begun to recover from over a century of extirpation from hunters and landowners. The first pack with pups was observed in 2008, presumably having immigrated from nearby states. Since then, the population has grown rapidly (table 17.1).

Here we will apply all of the tools developed so far to answer three questions:

1. Is there evidence of density-dependent population growth?
2. Is the process or observation error model better supported by the data?
3. Is this population likely to exhibit complex population dynamics (dampened oscillations, limit cycles, deterministic chaos)?

Introduction to Quantitative Ecology: Mathematical and Statistical Modelling for Beginners. Timothy E. Essington, Oxford University Press. © Timothy E. Essington 2021.
DOI: 10.1093/oso/9780192843470.003.0017

Table 17.1 *Counts of gray wolves in Washington State*

Year	Observed N
2008	5
2009	14
2010	19
2011	35
2012	51
2013	52
2014	68
2015	90
2016	115
2017	122
2018	126
2019	133

17.1   Fitting the observation error model

Recall from section 8.6 that an observation error model looks like this:

$$N_t = f(\theta) \tag{17.1}$$
$$N_{t,\text{obs}} \sim \text{Pois}(\lambda = N_t)$$

where $f(\theta)$ is a function of the model parameters. Here we are assuming that observations are random variables drawn from a Poisson probability function and that the population dynamics are *deterministic*. By substituting different functions for $f(\theta)$, we can estimate a density-independent or density-dependent model.

For instance, a density-independent function is

$$f(r, N_1) = \begin{cases} N_1 & \text{if } t = 1 \\ N_{t-1} + rN_{t-1} & \text{if } t > 1 \end{cases} \tag{17.2}$$

while a density-dependent function could be the logistic model:

$$f(r, N_1, K) = \begin{cases} N_1 & \text{if } t = 1 \\ N_{t-1} + rN_{t-1}\left(1 - \frac{N_{t-1}}{K}\right) & \text{if } t > 1 \end{cases} \tag{17.3}$$

Note here I'm using a slightly different notation for starting population size, calling it N_1 instead of N_0 as I usually do. The reason for this change will be clearer when we compare the observation model to the process error model.

17.1.1 Observation error model in spreadsheets

We first create a spreadsheet to run the density-independent model (equation 17.2), and adjust it to run the density-dependent model. List the data in your spreadsheet, starting in row 5, columns A and B, and place headers "Year" and "Observed Nt" in row 4 in columns A and B. As usual, place the parameters at the top, put the parameter value for r in cell B1, and put the initial population size N_1 in cell B2. Name these cells (though you can't name cell B1 "r," so use "rr" instead).

Follow the guidance you used to code the logistic population model in Excel (section 12.1) to generate N_t in column C, but use the density-independent model in place of the logistic model (table 17.2). Finally, in column D, calculate the negative log-likelihood for each observation (table 17.2).

Then, in cell D2, sum the negative log-likelihood over all of the observations:

=**SUM(D5:D16)**

Finally, use solver to minimize the negative log-likelihood by adjusting r and N_0.

Once your spreadsheet is setup like this, it is easy to modify it to include logistic density dependence (equation 17.3). First, you need to add the parameter K to the spreadsheet in cell B3 (and name the cell). Then, you replace the density-independent recursive equation in cells C6 through C16 with the logistic equation. For instance, in cell C6, you would write

Table 17.2 *Setting up your worksheet to estimate observation error parameters*

	A	B	C	D
1	rr	0.2		
2	N_1	15		=SUM(D5:D16)
3				
4	Year	Observed Number	Nt	Negative Log-Likelihood
5	2008	5	=N_1	=-LN(POISSON.DIST(B6,C6,FALSE))
6	2009	14	= C5 + C5 * rr	=-LN(POISSON.DIST(B6,C6,FALSE))
7	2010	19	...	...
8	2011	35	...	...
9	...	...	...	...

= C5 + rr * C5 * (1 - C5 / K)

and copy this formula down the C column. Once this is completed, you can rerun Solver, adding K to the list of parameters to adjust.

17.1.2   Observation error model in R

First load or list the data in an R script, and then write a function that receives a vector of parameters and the data, projects the model forward as we did in section 13.2, and returns the negative log-likelihood. For the density-independent model (equation 17.2), this might look like

```
nt.obs <- c(5,14,19,35,51,52,68,90,115,122,126,133)
obs.di.nll <- function(pars, nt.obs) {
  r <- pars[1]
  n_1 <- pars[2]
  nt <- rep(x = NA, times = length(nt.obs))
  nt[1]<-n_1
  n.loops <- length(nt.obs)
  for (i in 2:n.loops)  nt[i] <- nt[i-1]+ nt[i-1] *r
  nll <- -sum(dpois(nt.obs, lambda = nt, log = T))
  return(nll)
}
```

where nt.obs is the vector of wolf counts. Then set starting values for the model parameters, and run run optim so that you pass pass the vector nt.obs to the function

```
start.pars <- c(0.25,   10)
obs.di.fit <- optim(par = start.pars,
                    fn = obs.di.nll,
                    nt.obs= nt.obs,
                    method = "Nelder-Mead")
```

You might see warning messages that say

```
Warning messages:
1: In dpois(nt.obs, lambda = nt, log = T) : NaNs produced
```

This just means that, during the optimization routine, it tried a combination of parameters that made the N_t zero or negative. You can avoid that by using the log-link (section 16.1.2), where the vector pars consists of the log of the parameters

```
obs.di.nll <- function(pars, nt.obs) {
  r <-exp( pars[1])
  n_1 <- exp(pars[2])
  . . .
```

and you initialize the starting parameters as

```
start.pars <- log(c(0.25,  10))
```

This ensures that both r and N_0 are always positive. Just remember, when you use this method, the fitted parameter values that are returned are the log of the maximum likelihood parameter values.

To fit the logistic density-dependent model (equation 17.3), you only need to change one line of the code. You replace `for (i in 2:n.loops) nt[i] <- nt[i-1]+ nt[i-1] *r` with

```
for (i in 2:n.loops)  nt[i] <- nt[i-1]+
  nt[i-1] *r *(1 - nt[i-1] / K)
```

17.1.3   Evaluating fits of the observation error models

Figure 17.1 shows the data as points, and the fitted model deterministic trajectories as lines. Clearly, the density-independent model fails to capture the trend in the data. Note how it systematically overpredicts the number of wolves prior to 2011, underpredicts the number of wolves after 2012, and overpredicts again in 2018 and 2019. In comparison, the fit of the density-dependent model does not raise much alarm (figure 17.1). The data seem to be well predicted by the model, and there is no obvious trend or pattern in the observation errors. Not surprisingly, the negative log-likelihood of the density-

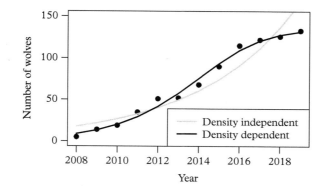

Figure 17.1 *Wolf counts through time (points) with best-fitting predictions from the density-independent (gray) and density-dependent (black) models as lines.*

independent model is relatively high compared to the density-dependent model (see section 17.2.3, table 17.4).

17.2   Fitting the process error model

Recall from section 8.7 that the process error model assumes that the observations are perfect, and deviations from the model predictions are due to randomness in the population process:

$$\mathbb{E}[N_t] = f(N_{t-1}, \theta)$$
$$N_t \sim \text{Pois}(\lambda = \mathbb{E}[N_t]) \tag{17.4}$$

where the expression $\mathbb{E}[N_t] = f(N_{t-1}, \theta)$ means that the expected population size in one time step depends on the observed population size in the previous time step N_{t-1} and the model parameters. This is basically the *stochastic* population model introduced in chapter 5, but with Poisson random variables instead of normal random variables. The main practical difference from the observation error model is that we calculate the expected value of N_t from the observed population size in the previous time step.

Common functions for density-dependent and density-independent process models are the two expressions

$$f(N_{t-1}, r) = N_{t-1} + rN_{t-1} \tag{17.5}$$
$$f(N_{t-1}, r, K) = N_{t-1} + rN_{t-1}\left(1 - \frac{N_{t-1}}{K}\right) \tag{17.6}$$

Because we want to compare the process error and observation error models, we have a small difficulty that we need to overcome. That is, if y is the number of years of data, the process error model uses $y - 1$ pairs of N_{t-1} versus N_t. Put more plainly, in the first year of data, there is no N_{t-1} available to us. By contrast, the observation error model calculates a likelihood for each of the y years. Remember that, to use AIC, the models have to be fit to the same set of data (chapter 9). For that reason, we'll set up our process error model to use all y years in the likelihood. Namely, instead of estimating the initial population size to apply in 2008 as we did above, we will estimate the initial population size as the number of (unobserved) wolves in 2007 (N_0). We then use that estimated population size to generate a prediction of the expected number of wolves for 2008, the first year of data.

With this modification, the function for the density-independent model is

$$f(N_{t-1}, r, N_0) = \begin{cases} N_0 + rN_0 & \text{if } t = 1 \\ N_{t-1} + rN_{t-1} & \text{if } t > 1 \end{cases}$$

and that for the density-dependent model is

$$
f(N_{t-1}, r, N_0, K) = \begin{cases} N_0 + rN_0\left(1 - \frac{N_0}{K}\right) & \text{if } t = 1 \\ N_{t-1} + rN_{t-1}\left(1 - \frac{N_{t-1}}{K}\right) & \text{if } t > 1 \end{cases}
$$

Once we have the expected population sizes for each year, we calculate the negative log-likelihood for all of the remaining observations by applying equation 17.4.

17.2.1   Process error model in spreadsheets

We will adapt the spreadsheet developed above (section 17.1.1) so that it uses the process error model. As before, we start with the density-independent model and then introduce the new parameter N_0 as the population size in 2007. Once that parameter is added, we only need to modify column C, where, instead of generating the deterministic N_t, it will instead calculate the expected N_t based on the observed population size for the previous time step (table 17.3). Once your spreadsheet is setup in this manner, use Solver to estimate the starting population size and population growth rate.

To make this a density-dependent model, we adjust cell C5 to be

=N_0 + N_0 * rr * (1 - N_0 / K)

and then adjust cells C6 and downwards to be

=B5 + B5 * rr * (1 - B5 / K)

17.2.2   Process error model in R

We need to write new functions to apply the process error model. The density-independent process error model function might look like

Table 17.3 *Setting up your worksheet to estimate process error parameters*

	A	B	C	D
1	rr	0.2		
2	N_0	15		
3				
4	Year	Observed Number	E[Nt]	Negative Log-Likelihood
5	2008	5	=N_0 + N_0 * rr	=-LN(POISSON.DIST(B6,C6,FALSE))
6	2009	14	= B5 + B5 * rr	=-LN(POISSON.DIST(B6,C6,FALSE))
7	2010	19	...	...
8	2011	35	...	...
9	...	...	...	...

```
process.di.nll <- function(pars, nt) {
  r <- pars[1]
  n_0 <- pars[2]
  nt.hat <- rep(NA, length(nt))
  nt.hat[1] <- n_0  + n_0 * r
  n.loops <- length(nt)
  for (i in 2:n.loops) {
    nt.hat[i] <- nt[i-1] + nt[i-1] * r
  }
  return(-sum(dpois(x = nt, lambda = nt.hat, log = TRUE)))
}
```

where the vector nt.hat is giving the expected population size for each year.

You might choose to use the log-link trick to remove the annoying warning messages, but I'll keep the code simpler here for demonstration purposes. You then use optim as follows:

```
start.pars <- c(0.2, 35)
process.di.fit <- optim(par = start.pars,
                        fn = process.di.nll,
                        nt= nt.obs,
                        method = "BFGS")
```

Finally, to create a density-dependent model, you only need to change two lines in the above function. Replace nt.hat[1] <- n_0 + n_0 * r with

```
nt.hat[1]  <- n_0  + n_0 * r * (1 - n_0 / K)
```

and replace nt.hat[i] <- nt[i-1] + nt[i-1] * r with

```
nt.hat[i]  <- nt[i-1] +
      nt[i-1] * r * (1 - nt[i-1] / K)
```

17.2.3   Evaluating fits of the process error models

Figure 17.2 shows the data as pairs of N_{t-1} and N_t, with the expected N_t from the best-fitting parameter values for each model shown as lines. Once again, the density-independent model fit looks disappointing. The model generally underpredicts the observed wolf counts when N_{t-1} is less than about 100, and overpredicts otherwise. As before, the density-dependent model appears to provide a much better fit to the data (figure 17.2). Once again, the negative log-likelihood is substantially lower

Table 17.4 *Parameter estimates, negative log-likelihood, AIC, and ΔAIC for each model*

Error type	Model type	r	N_1 or N_0	K	Negative log-likelihood	AIC	ΔAIC
Observation	Density independent	0.23	17.9	—	54.9	113.8	32.2
Observation	Density dependent	0.54	8.9	135.5	37.8	81.6	0.0
Process	Density independent	0.18	4.2	—	47.2	98.4	16.8
Process	Density dependent	0.59	3.3	132.8	40.0	86.0	4.4

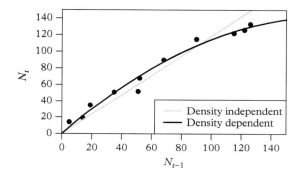

Figure 17.2 *Wolf population counts in time $t-1$ versus time t, with the best fits of the density-independent (gray) and density-dependent (black) models as lines.*

for the density-dependent model than for the density-independent model (table 17.4), but the difference is not as large as it was for the observation error model.

17.3   Parameter estimates and model selection

The parameter estimates were similar for different error types, but quite different for each model type. This is not uncommon, especially r which has a different meaning for the two model types—for the density-dependent models, it is the maximum population growth rate, while, for the density-independent models, it is the average population growth rate.

The ΔAIC values indicate substantial support for the density-dependent models. The top two models are both density-dependent models, and the density-independent models have ΔAIC well beyond the levels where we would continue to give them consideration. The data also give stronger support to the observation error density-dependent model over the process error density-dependent model. The difference in AIC is not so large that we would outright dismiss the process error model, however.

17.4 Can this population exhibit complex population dynamics?

The model selection indicated that the density-dependent deterministic (observation error) model had the strongest support. We saw in section 2.5 that models will exhibit monotonic increase to carrying capacity as long as r is less than 1. The maximum likelihood estimate of r is 0.54, but is $r \geq 1$ plausible? We can address that by generating a confidence interval for r, using likelihood profiles. This is challenging to do in spreadsheets, because we have three parameters: one of interest and two nuisance parameters. We could only do this in spreadsheets if we were comfortable fixing the estimated starting population size at the maximum likelihood estimate and then doing a likelihood profile over all K to get the upper and lower bounds for r. We would use the data table to loop over candidate values of r and K, as described in section 16.3.1. You might try values of r ranging from 0.4 to 1.1 in increments of 0.025, and values of K ranging from 110 to 160 in increments of 2.

In R, we can account for uncertainty in both carrying capacity and starting population size. First, we create a function that accepts a vector `pars` that contains only K and N_1, plus a single fixed value of r and the observed population sizes:

```
profile.nll <- function(pars, r, nt.obs) {
  K <- pars[1]
  n_1 <- pars[2]
  nt <- rep(x = NA, times = length(nt.obs))
  nt[1]<-n_1
  n.loops <- length(nt.obs)
  for (i in 2:n.loops)  nt[i] <- nt[i-1] +
    nt[i-1] * r * (1 - nt[i-1] / K)

  # calculate NLL assuming poisson observation errors
  nll <- -sum(dpois(nt.obs, lambda = nt, log = T))
  return(nll)
}
```

We then create a list of values of r, cycle through them, and, for each, calculate the profile negative log-likelihood by optimizing the function `profile.nll`:

```
rlist <- seq(from = 0.3, to =1.1, length.out = 100)
r.profile <- rep(NA, times = length(rlist))
for (i in 1:length(rlist)) {
  profile.fit <- optim(par = c(135, 8.9),
                       fn = profile.nll,
                       r = rlist[i],
                       nt.obs = nt.obs,
                       method = "Nelder-Mead")
  r.profile[i] <- profile.fit$value
  }
```

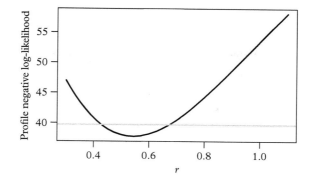

Figure 17.3 *Likelihood profile for* r *for the observation error density-dependent model, fit to the Washington wolf data.*

The plot of the likelihood profile certainly does not support the hypothesis that r is greater than 1 (figure 17.3). We can be more deliberate by applying the same procedures described in section 16.3.2 to get the upper and lower confidence bounds. The procedure yields a confidence interval of $(0.43, 0.67)$, far below 1.0. We therefore conclude that it is highly unlikely that the intrinsic rate of population growth of this population could lead to dampened oscillations, limit cycles, or deterministic chaos.

Describe, explain, and interpret

Describe: The two density-dependent models had considerably more support than the density-independent models. The observation error model had somewhat more support than the process error model. The confidence interval for r did not include 1, making complex population dynamics like dampened oscillations, limit cycles, or deterministic chaos highly unlikely.

Explain: The data, plotted as either N_t versus time or N_t versus N_{t-1} show clear evidence that rates or population growth were slower at the end of the time series, when population levels were at their highest. The slightly greater support of the observation error over the process error model likely is due to the 2012 and 2013 data points. Under the observation error model, neither of these are terribly unusual, with the first being slightly above the predicted N_t, and the last being slightly below the predicted N_t. Under the process error model, the model has to reconcile the large jump in population size from 2011 to 2012, and then the very low change in population size from 2012 to 2013 (the latter appearing as the most unusual data point in figure 17.2). We also do not expect medium-sized mammals to have rates of population growth that exceed $r > 1$, based on simple life-history properties.

continued

Interpret: The population is clearly showing reduced population growth at high population densities. But we should take care not to attribute this solely to resource limitation. That is, other factors might be important in the latter part of the time series. For example, in 2019 alone, nine wolves were removed because they had killed livestock, six were legally hunted by tribes, and four were killed by landowners to protect their livestock or for personal safety. Thus, we might interpret the population dynamic as reflecting the carrying capacity of the linked social-human system, rather than the carrying capacity of the natural system itself.

18

Next Steps

Congratulations, you've made it all of the way through! In Part 1, you learned to apply the "describe, explain, and interpret" framework to make inferences about the real world from mathematical models. You've seen how models with one or more state variables can be constructed, and how random stochastic events can be considered. In Part 2, you learned that one can pose models as hypotheses, asking whether the data supports one model, function, or set of parameter values over another. Finally, in Part 3, you gained several technical tools for mathematical and statistical modeling. You're well on your way to building and using models for your research questions while also gaining the knowledge base to critically evaluate your peers' quantitative research.

18.1 Reality check

For the purposes of learning the fundamentals, the examples posed here were relatively tractable. In Part 1, we mostly focused on population or community models, illustrating commonalities and differences among models and developing your ability to evaluate them. In Part 2, we often had one or maybe two parameters to estimate, and we only explored how to fit to unstructured dynamic models.

Of course, the real world is full of more complex dimensions that this book didn't cover. This should not be a surprise, really; each one of these chapters has one or more entire books devoted to the topics covered in them! On the mathematical ecology side of things, here are some models that you might encounter in your field:

- agent-based models (modeling individuals or groups of individuals)
- spatially explicit models
- population genetics, or eco-evolutionary models
- eco-physiological models
- models of life-history trade-offs and evolution of traits
- topological models of systems
- more complex analyses of extinction risk that account for density dependence, spatial heterogeneity, and catastrophes

Introduction to Quantitative Ecology: Mathematical and Statistical Modelling for Beginners. Timothy E. Essington, Oxford University Press. © Timothy E. Essington 2021.
DOI: 10.1093/oso/9780192843470.003.0018

- optimal foraging models
- diseases and epidemics
- models of individual growth
- full ecosystem models
- size-based models, integro-difference models, and other ways of modeling distinctions among individuals based on traits
- delay models (where the value of state variables at different times govern rates)

On the statistical ecology side of things, you will find all manner of additional complexities, such as the following:

- generalized linear models and generalized linear mixed-effects models
- state-space models to account for both observation and process error
- time series models
- hierarchical models
- spatiotemporal models
- empirical dynamic models and nonlinear forecasting
- machine learning, such as Gaussian processes, random forests, and neural networks
- more advanced Bayesian and model selection methods

18.2   Learn by doing

I hope that last section didn't overwhelm you. If it did, take a moment to consider a few things. One, virtually no one becomes expert at all of these topics. Two, the point was to show the types of things you might see in more advanced textbooks and scientific literature, and the opportunities they present to expand your skills and knowledge. Three—the most important thing—is that mastering any of those specialized skills will build on the foundational concepts and skills you've already mastered here. Using the "describe, explain, and interpret" framework will serve you well in *any* quantitative context. Systematically understanding how and why we use models will open your eyes to new ways you might conduct your own research. The statistical methods have, at their core, the same basic principles that we covered here—they merely expand to cover additional levels of complexity that the real world throws at us.

What I'm trying to say is that becoming a card-carrying quantitative ecologist is a process. This book hopefully started you on this journey. Where you go next is up to you. Maybe you are perfectly happy with where you stand right now. You have some new ideas to apply to your own research, and you have a sharper eye to evaluate other people's quantitative work. Great!

Or maybe this book ignites an interest in learning more methods and more models. Great! Probably the best way to proceed is to start applying these methods to your own research. And you only need a model that is useful, not one that is sophisticated, elegant, or complicated.

18.2.1   **True story**

My very first first-authored paper had a little model in it, all programmed in a now-defunct spreadsheet (Quatro Pro, run in MS-DOS). I had just taken an introductory modeling class as a graduate student, and my research was looking at how frequently a certain species of fish tended to dig their nests on top of previously dug nests. I wanted to know whether the frequency of this behavior could be explained by habitat availability. So, I modeled the process of nest site selection as a Monte Carlo simulation and asked how often would I see the observed frequency if fish were selecting nest sites purely on habitat features. I found that it was vanishingly unlikely—fish reused nest sites far more often than expected by chance. I now know, looking back, that there were more elegant ways to answer this question. But who cares—the model I used did the job!

Bibliography

Abrams, P. A. 2009. "When Does Greater Mortality Increase Population Size? The Long History and Diverse Mechanisms Underlying the Hydra Effect." *Ecology Letters* **12**(5), 462–74.

Aho, K., Derryberry, D., and Peterson, T. 2014. "Model Selection for Ecologists: The Worldviews of AIC and BIC." *Ecology* **95**(3), 631–36.

Akaike, H. 1974. "A New Look at the Statistical Model Identification." *IEEE Transactions on Automatic Control* **19**(6), 716–23.

Aldrich, J. 1997. "R. A. Fisher and the Making of Maximum Likelihood 1912–1922." *Statistical Science* **12**(3), 162–76.

Allee, W. 1927. "Animal Aggregations." *Quarterly Review of Biology* **2**(3), 367–98.

Anderson, D. R., Burnham, K. P., and Thompson, W. L. 2000. "Null Hypothesis Testing: Problems, Prevalence, and an Alternative." *Journal Of Wildlife Management* **64**(4), 912–23.

Anderson, S. C., Branch, T. A., Cooper, A. B., and Dulvy, N. K. 2017. "Black-Swan Events in Animal Populations." *Proceedings of the National Academy of Sciences* **114**(12), 3252–57.

Arcese, P., and Smith, J. N. M. 1988. "Effects of Population Density and Supplemental Food on Reproduction in Song Sparrows." *Journal of Animal Ecology* **57**(1), 119–36.

Baird, D., McGlade, J. M., and Ulanowicz, R. E. 1991. "The Comparative Ecology of Six Marine Ecosystems." *Philosophical Transactions of the Royal Society of London Series B: Biological Sciences* **333**(1266), 15–29.

Baskett, M. L., Weitz, J. S., and Levin, S. A. 2007. "The Evolution of Dispersal in Reserve Networks." *American Naturalist* **170**(1), 59–78.

Bem, D. J. 2011. "Feeling the Future: Experimental Evidence for Anomalous Retroactive Influences on Cognition and Affect." *Journal of Personality and Social Psychology* **100**(3), 407–25.

Bolker, B. M., Brooks, M., Clark, C., Geange, S., Paulsen, J., Stevens, M., and White, J.-S. 2008. "Generalized Linear Mixed Models: A Practical Guide for Ecology and Evolution." *Trends in Ecology & Evolution* **24**(3), 127–35.

Box, G. 1979. "Robustness in the Strategy of Scientific Model Building." Chap. 7 in *Robustness in Statistics*. New York, NY: Academic Press.

Brook, B. W., O'Grady, J. J., Chapman, A. P., Burgman, M. A., Akçakaya, H. R. and Frankham, R. 2000. "Predictive Accuracy of Population Viability Analysis in Conservation Biology." *Nature* **404**(6776), 385–87.

Burnett, K. M., D'Evelyn, S., Kaiser, B. A., Nantamanasikarn, P., and Roumasset, J. A. 2008. "Beyond the Lamppost: Optimal Prevention and Control of the Brown Tree Snake in Hawaii." *Ecological Economics* **67**(1), 66–74.

Burnham, K. P., and Anderson, D. R. 1998. *Model Selection and Inference: A Practical Information-Theoretic Approach*. New York, NY: Springer-Verlag.

Burnham, K. P., and Anderson, D. R. 2014. "P Values Are Only an Index to Evidence: 20th- vs. 21st-Century Statistical Science." *Ecology* **95**(3), 627–30.

Caswell, H. 2000a. *Matrix Population Models: Construction, Analysis and Interpretation*. 2nd ed. Sunderland, MA: Sinauer Associates.

Caswell, H. 2000b. "Prospective and Retrospective Perturbation Analyses: Their Roles in Conservation Biology." *Ecology* **81**(3), 619.

Chamberlain, T. (1890) 1965. "The Method of Multiple Working Hypotheses." *Science* 148(3671), 754–59.

Clark, J. S. 2004. "Why Environmental Scientists Are Becoming Bayesians." *Ecology Letters* 8(1), 2–14.

Crouse, D. T., Crowder, L. B., and Caswell, H. 1987. "A Stage-Based Population-Model for Loggerhead Sea-Turtles and Implications for Conservation." *Ecology* 68(5), 1412–23.

DeAngelis, D. L. 1992. *Dynamics of Nutrient Cycling and Food Webs*. Dordrecht: Springer.

Dennis, B. 1996. "Discussion: Should Ecologists Become Bayesians?." *Ecological Applications* 6(4), 1095–103.

Dennis, B., Munholland, P. L., and Scott, J. M. 1991. "Estimation of Growth and Extinction Parameters for Endangered Species." *Ecological Monographs* 61(2), 115–43.

Dunne, J. A., Williams, R. J., and Martinez, N. D. 2002. "Network Structure and Biodiversity Loss in Food Webs: Robustness Increases with Connectance." *Ecology Letters* 5(4), 558–67.

Edwards, A. W. F. 1974. "The History of Likelihood." *International Statistical Review/Revue Internationale de Statistique* 42(1), 9–15.

Ellner, S. P., and Rees, M. 2006. "Integral Projection Models for Species with Complex Demography." *The American Naturalist* 167(3), 410–28.

Essington, T., Baskett, M. L., Sanchirico, J., and Walters, C. 2015. "A Novel Model of Predator–Prey Interactions Reveals the Sensitivity of Forage Fish: Piscivore Fishery Trade-offs to Ecological Conditions." *ICES Journal of Marine Science* 72, 1349–58.

Essington, T., Ward, E. J., Francis, T. B., Greene, C., Kuehne, L., and Lowry, D. 2021. "Historical Reconstruction of the Puget Sound (USA) Groundfish Community." *Marine Ecology Progress Series* 657, 173–89.

Ferriss, B., and Essington, T. 2014. "Does Trophic Structure Dictate Mercury Concentrations in Top Predators? A Comparative Analysis of Pelagic Food Webs in the Pacific Ocean." *Ecological Modelling* 278, 18–28.

Fisher, A., R. 1912. "On an Absolute Criterion for Fitting Frequency Curves." *Messenger of Mathematics* 41, 155–160.

Forrester, G. E. 1995. "Strong Density-Dependent Survival and Recruitment Regulate the Abundance of a Coral Reef Fish." *Oecologia* 103(3), 275–82.

Fulton, E. A., Smith, A. D., Smith, D. C., and Johnson, P. 2014. "An Integrated Approach is Needed for Ecosystem Based Fisheries Management: Insights from Ecosystem-Level Management Strategy Evaluation." *PLOS ONE* 9(1), e84242.

Gelman, A., Carlin, J., Stern, H., and Rubin, D. 1995. *Bayesian Data Analysis*. London: Chapman & Hall.

Gelman, A., and Hill, J. 2007. *Data Analysis Using Regression and Multilevel/Hierarchical models*. New York, NY: Cambridge University Press, New York.

Gerber, L. R., and Heppell, S. S. 2004. "The Use of Demographic Sensitivity Analysis in Marine Species Conservation Planning." *Biological Conservation* 120(1), 121–28.

Gilpin, M., and Soule, M. E. 1986. "Minimum Viable Populations: Processes of Species Extinctions." In *Conservation Biology: The Science of Scarcity and Diversity*, edited by M. E. Soule, 19–34. Sunderland, MA: Sinauer Associates.

Green, L. E., and Peloquin, J. E. 2008. "Acute Toxicity of Acidity in Larvae and Adults of Four Stream Salamander Species (Plethodontidae)." *Environmental Toxicology and Chemistry* 27(11), 2361–67.

Harte, J. 1985. *Consider a Spherical Cow: A Course in Environmental Problem Solving*. Mill Valley, CA: University Science Books.

Heyde, C., and Cohen, J. 1985. "Confidence Intervals for Demographic Projections Based on Products of Random Matrices." *Theoretical Population Biology* 27(2), 120–53.

Hilborn, R., and Mangel, M. 1997. *The Ecological Detective.* Princeton NJ: Princeton University Press.

Hilborn, R., and Walters, C. 1992. *Quantitative Fisheries Stock Assessment.* Boston, MA: Kluwer Academic Publishers.

Hobbs, N. T., and Hooten, M. 2015. *Bayesian Models: A Primer for Ecologists.* Princeton, NJ: Princeton University Press.

Holmes, E. E. 2004. "Beyond Theory to Application and Evaluation: Diffusion Approximations for Population Viability Analysis." *Ecological Applications* 14(4), 1272–93.

Hooten, M., and Hobbs, N. T. 2015. "A Guide to Bayesian Model Selection for Ecologists." *Ecological Monographs* 85(1), 23–28.

Johnson, J. B., and Omland, K. S. 2004. "Model Selection in Ecology and Evolution." *Trends in Ecology & Evolution* 19(2), 101–8.

Kendeigh, S. 1982. *Bird Populations in East Central Illinois: Fluctuations, Variation and Development over a Half-Century.* Illinois Biological Monographs, vol. 52. Champaign, IL: University of Illinois Press.

Kodric-Brown, A., and Brown, J. H. 1978. "Influence of Economics, Interspecific Competition, and Sexual Dimorphism on Territoriality of Migrant Rufous Hummingbirds." *Ecology* 59(2), 285–96.

Konishi, S., and Kitagawa, G. 2008. "Information Criterion." Chap. 3 in *Information Criteria and Statistical Modeling.* Springer Series in Statistics. New York, NY: Springer, New York, NY.

Kullback, S., and Leibler, R. A. 1951. "On Information and Sufficiency." *The Annals of Mathematical Statistics* 22(1), 79–86.

Levins, R. 1966. "The Strategy of Model Building in Population Biology." *American Scientist* 54(4), 421–31.

Lewis, M., and Karieva, P. 1993. "Allee Dynamics and the Spread of Invading Organisms." *Theoretical Population Biology* 43(2), 141–58.

Liermann, M., and Hilborn, R. 1997. "Depensation in Fish Stocks: A Hierarchic Bayesian Meta-analysis." *Canadian Journal of Fisheries and Aquatic Sciences* 54(9), 1976–84.

Mangel, M. 2006. *The Theoretical Biologist's Toolbox.* Cambridge: Cambridge University Press.

Martien, K. K., Taylor, B. L., Slooten, E., and Dawson, S. 1999. "A Sensitivity Analysis to Guide Research and Management for Hector's Dolphin." *Biological Conservation* 90(3), 183–91.

May, R. M. 1973. *Stability and Complexity in Model Ecosystems.* Princeton, NJ: Princeton University Press.

May, R. M. 1976. "Simple Mathematical Models with Very Complicated Dynamics." *Nature* 261(5560), 459–67.

Monnahan, C. C., Thorson, J. T., and Branch, T. A. 2017. "Faster Estimation of Bayesian Models in Ecology Using Hamiltonian Monte Carlo." *Methods in Ecology and Evolution* 8(3), 339–48.

Morris, W., and Doak, D. 2002. *Quantitative Conservation Biology.* Sunderland, MA: Sinauer Associates.

Murakami, H. 1991. *Hard Boiled Wonderland and the End of the World.* New York, NY: Vintage Press.

Murtaugh, P. A. 2014. "In Defense of *P* Values." *Ecology* 95(3), 611–17.

Nåsell, I. 2001. "Extinction and Quasi-stationarity in the Verhulst Logistic Model." *Journal of Theoretical Biology* 211(1), 11–27.

Norden, R. H. 1982. "On the Distribution of the Time to Extinction in the Stochastic Logistic Population Model." *Advances in Applied Probability* **14**(4), 687–708.

Otto, S. P., and Day, T. 2007. *A Biologist's Guide to Mathematical Modeling in Ecology and Evolution.* Princeton, NJ: Princeton University Press.

Pelton, E. M., Schultz, C. B., Jepsen, S. J., Black, S. H., and Crone, E. E. 2019. "Western Monarch Population Plummets: Status, Probable Causes, and Recommended Conservation Actions." *Frontiers in Ecology and Evolution*, 258.

Pickett, S., Kolasa, J., and Jones, C. 1994. *Ecological Understanding.* San Diego, CA: Academic Press.

Plagányi, E. E., van Putten, I., Hutton, T., Deng, R. A., Dennis, D., Pascoe, S., Skewes, T., and Campbell, R. A. 2013. "Integrating Indigenous Livelihood and Lifestyle Objectives in Managing a Natural Resource." *Proceedings of the National Academy of Sciences of the United States of America* **110**(9), 3639–44.

Platt, J. R. 1964. "Strong Inference: Certain Systematic Methods of Scientific Thinking May Produce Much More Rapid Progress Than Others." *Science* **146**(3642), 347–53.

Polacheck, T., Hilborn, R., and Punt, E. A. 1993. "Fitting Surplus Production Models: Comparing Methods and Measuring Uncertainty." *Canadian Journal of Fisheries and Aquatic Sciences* **50**(12), 2594–607.

Pollard, E., Lakhani, K. H., and Rothery, P. 1987. "The Detection of Density-Dependence from a Series of Annual Censuses." *Ecology* **68**(6), 2046–55.

Price, P. W., Bouton, C. E., Gross, P., McPheron, B. A., Thompson, J. N., and Weis, A. E. 1980. "Interactions Among Three Trophic Levels: Influence of Plants on Interactions between Insect Herbivores and Natural Enemies." *Annual Review of Ecology and Systematics* **11**(1), 41–65.

Schnute, J. T. 1994. "A General Framework for Developing Sequential Fisheries Models." *Canadian Journal of Fisheries and Aquatic Sciences* **51**(8), 1676–88.

Sinclair, A. R. E., and Pech, R. P. 1996. "Density Dependence, Stochasticity, Compensation and Predator Regulation." *Oikos* **75**(2), 164–73.

Skalski, G., and Gilliam, J. 2001. "Functional Responses with Predator Interference: Viable Alternatives to the Holling Type II Model." *Ecology* **82**(11), 3083–92.

Starfield, A. 1997. "A pragmatic Approach to Modeling for Wildlife Management." *Journal of Wildlife Management* **61**(2), 261–70.

Stephens, P. A., Buskirk, S. W., and Martinez del Rio, C. 2007. "Inference in Ecology and Evolution." *Trends in Ecology & Evolution* **22**(4), 192–97.

Stephens, P. A., Buskirk, S. W., Hayward, G. D., and Martinez del Rio, C. 2005. "Information Theory and Hypothesis Testing: A Call for Pluralism." *Journal of Applied Ecology* **42**(1), 4–12.

Taleb, N. N. 2010. *The Black Swan: The Impact of the Highly Improbable.* New York, NY: Random House.

Thiessen, E. D., Hill, E. A., and Saffran, J. R. 2005. "Infant-Directed Speech Facilitates Word Segmentation." *Infancy* **7**(1), 53–71.

Tuljapurkar, S. and Orzack, S. 1980. "Population Dynamics in Variable Environments. I. Long-Run Growth Rates and Extinction." *Theoretical Population Biology* **18**(3), 314–42.

Tversky, A., and Kahneman, D. 1983. "Extensional versus Intuitive reasoning: The Conjunction Fallacy in Probability Judgment." *Psychological Review* **90**(4), 293–315.

Valpine, P. D., and Hastings, A. 2002. "Fitting Population Models Incorporating Process Noise and Observation Error." *Ecological Monographs* **72**(1), 57–76.

van Beest, F. M., Kindt-Larsen, L., Bastardie, F., Bartolino, V., and Nabe-Nielsen, J. 2017. "Predicting the Population-Level Impact of Mitigating Harbor Porpoise Bycatch with Pingers and Time-Area Fishing Closures." *Ecosphere* **8**(4), e01785.

Varley, G. 1949. "Population Changes in German Forest Pests." *Journal of Animal Ecology* **18**(1), 117–22.

Wagenmakers, E.-J., Wetzels, R., Borsboom, D., and van der Maas, H. L. J. 2011. "Why Psychologists Must Change the Way They Analyze Their Data: The Case of Psi: Comment on Bem (2011)." *Journal of Personality and Social Psychology* **100**(3), 426–32.

Walters, C. 1986. *Adaptive Management of Renewable Resources*. New York, NY: Macmillan Publisers.

Ward, E. J., Ford, M. J., Pope, R. G., Ford, J. K., Velez-Espino, L. A., Parken, C. K., Lavoy, L. W., Handon, M. B., and Balcom, K. C. 2013. *Estimating the Impacts of Chinook Salmon Abundances and Prey Removal by Ocean Fishing on Southern Resident Killer Whale Population Dynamics*. NOAA Technical Memorandum NMFS-NWFSC-123. Washington, DC: U.S. Department of Commerce National Oceanic and Atmospheric Administration.

Werner, P. A., and Caswell, H. 1977. "Population Growth Rates and Age versus Stage-Distribution Models for Teasel (*Dipsacus sylvestris* Huds.)." *Ecology* **59**(1), 53–66.

Index